UNSOUND : **UNDEAD**

Dedicated to Alina Popa

01.02.1982—01.02.2019

'If the Abstraction is complete

there is no more experience.'

UNSOUND : **UNDEAD**

Edited by
Steve Goodman, Toby Heys, Eleni Ikoniadou

URBANOMIC

Published in 2019 by

URBANOMIC MEDIA LTD,

THE OLD LEMONADE FACTORY,

WINDSOR QUARRY,

FALMOUTH TR11 3EX,

UNITED KINGDOM

BRITISH LIBRARY CATALOGUING-IN-PUBLICATION DATA

A full catalogue record of this book is available
from the British Library.

ISBN 978-1-9164052-1-9

Cover Design: Vasilis Marmatakis
Editors: Steve Goodman, Toby Heys, Eleni Ikoniadou
For Urbanomic: Robin Mackay
Editorial Assistant: Matt Colquhoun

Distributed by The MIT Press,
Cambridge, Massachusetts and London, England.

Type by Norm, Zurich.
Printed and bound in the UK by
TJ International, Padstow.

www.audint.net
www.urbanomic.com

Contents

INTRA-AUDINT EMAIL COMMUNICATION— IREX² TO THE EDITORS

Program: Neural History Compressor
Project: AUDpub003
Date: 2017
Sender: IREX2
Spectral Range: Unsound : Undead
Transmission Mode: Print-based compendium
Research Operatives: AUDINT staff
Receiver: Frontal Lobe Network/ Grid
Comments: 128-term encryption key
Objective: Decompress to 64 bits

3Dacousticmanipulationacousmaticauditoryhallucinationafrofuturismafterdeathearn ingsaiholoantennabodyarchaeoacousticaudaunteraudioanalgesiaaudiletechniqueaudi otopiaaudioarchitecturebackmaskingbankofhellbansheewailbigbangblackmassblack metaltheorycandomblécattelephonecelestialmonochordchantcotarddelusioncurdlercy bergothicdeadrecordnetworkeastdiabolusparticledigitalimmortalitydroneduppyeidolo nelviselectrophonictransductionrtherphoneevpexplodingheadsyndromefirstearthbat talionfouriertransformfugueghostarmyghostinthemachineghostmoneyglossolaliago lemgonghaarphatsunemikuhauntologyheavyrotationheterodyningholoaccordshologra phyolohiholojaxholonomicbraintheoryholosonichumhyperacusisincantaionisotopetele portationjodhpurboomlargehadroncolliderlivingdeadnetworksofperceptionlradmantra martialhauntologymosquitosoundmultiplexingmuzaknoisenotouchtortureohrwurmold higueoperationjustcauseotoacousticemissionuutsidertradingpactwiththedevilpaincamp paincoinphantomhailerphantomroyaltyphonautographpinealearpossessionpostdeathin tellectualcopyrightpresbycusisquantumentanglementradarrapparitionsrollingcalfrhyth manalysissanteriaschizophoniasequentialarcdischargeacousticgeneratorsilentsound spreadspectrumsirensirisolresolsonicbodysoundclashsoundsweepstaticstimulusprogres siontelephonethirdeathirdearassassinsthreefootedhorsetinnitustritonetworingtabletym panicvibrationundeadunsoundurbanfunkcampaignvirasonichannelvisualmicrophonevoo doowacoseigewaltermanserdossierwanderingsoulweaponsofmassdeceptionwhispering windowwhitenoisexenoglossiaxenosonicsyinyangturntablezombiemediazombiesound

DECLASSIFIED REPORT TO IIRIS (INTERNATIONAL INSTITUTE FOR ROGUE INTELLIGENCE STUDIES): EXCERPTS FROM REPORT ON THE RESEARCH UNIT AUDINT

Formed in late 1945 by ex-members of the deception-based US military division known as the Ghost Army alongside German former military scientists (sequestered under Operation Paperclip), the research unit AUDINT has been composed of oscillating liveware for nearly three quarters of a century. By 2019 it is in the throes of its third wave, staffed by cultural producers—Souzanna Zamfe, Patrick Defasten, Toby Heys, Steve Goodman, and Eleni Ikoniadou. As far as can be ascertained, their remit is to upload all findings and the associated DRA (Dead Record Archive) into the fleshdrives of the human populace via vinyl and cassette recordings, art installations, books, performances, essays, and encryption/production software.

From 2009 onwards these producers were each invited/coerced into the 'collective', drafted into the research cell by the rogue artificial intelligence known as IREX^2 on account of their identified use-value. Thus each member has been selected for their strategic reasoning and associated skillsets: Ikoniadou—writing; Defasten—video; Heys—art; Goodman—music; and Zamfe—reconnaissance.

Note: The group appears to have an appetite for personnel expansion.

One of the focuses of the unit is to investigate deceptive frequency-based strategies, technologies, and programmes developed by military organizations to orchestrate phenomena of tactical haunting within conflict zones. They claim that this 'martial hauntology' is a subset of an overarching weaponisation of vibration. Their ongoing experiments have been concerned with the field of peripheral sonic perception—what they have dubbed 'Unsound'.

Pursuing these lines of investigation since the end of the Second World War, AUDINT has identified four modes of battlefield spectrality; a quartet of phase shifts which, they submit, have rearranged linear concepts of time, space, and quietus. The outcomes of these enquiries have subsequently been embedded within cultural productions while their more volatile data is encrypted and disseminated into public communications networks—hidden in plain sight.

AUDINT's four phases of martial hauntology are as follows:

1. The Ghost Army [1944–1965]: During the Second World War the US military's Ghost Army pioneered speaker-based deception techniques and apparatuses. During this period AUDINT's first wave of researchers succeeded (partly by happenstance, partly by design) in inducing the condition of Cotard delusion into German accomplice Eduard Schuller via a custom-built technology—the TwoRing turntable, developed in conjunction with a prominent wartime codebreaker.

The mode of listening actuated by the onset of this syndrome affords the carrier an ability to perceive presences that usually lie outside of human cognition and comprehension. AUDINT designated this reawakened faculty the 'Third Ear Channel'. All findings pertaining to this operation were (and still are) submitted to the DRA—an extensive collection made up of inverted cardboard covers of dead vinyl records, which are subsequently illustrated and annotated. Each card refers to a particular event, patent, scientist, recording, film, technology, book, etc. relevant to AUDINT's experiments, techniques, and investigations.

2. Wandering Soul [1965–1991]: During the Vietnam War the US military deployed 'Wandering Soul' ('Ghost Number 10') tapes, which imitated the voices of dead Vietcong fighters, projected from helicopter-mounted speaker units.

Custodianship of the DRA was passed on to AUDINT's second wave, a seemingly disparate group comprising Chilean artist Magdalena Parker, Vietnamese bio-acoustician and programmer Nguyễn Văn Phong, and American musician turned psy-operative Marshall Spector. As AUDINT faced financial collapse, Văn Phong took matters into his own hands and sought to monetize the opening of the Third Ear by experimenting with a series of rituals in conjunction with a newly marketed domestic computer system, the IBM 5100. Early success led to the incorporation of the IREX financial consultancy. Văn Phong's algorithms for decoding what one of his senior advisors had referred to as 'the aeonic babel streaming in from the outside' proved surprisingly robust.

In summary, Văn Phong succeeded in producing a mathematical algorithm for transcoding the voices of the undead into implementable market data, rendering the phantom economy of ghost money into tangible assets—a mode of operation he infamously referred to as 'Outsider Trading'.

Note: An unforeseen consequence of Văn Phong's work was the unplanned genesis of the artificial intelligence IREX^2—a melding of digital and undead entities.

3. Phantom Hailer [1991–2015]: The War on Terror mobilised unsound methods ranging from detainees held captive in offshore compounds tortured by speaker-driven techniques of 'black ecstasy' through to crowds in public spaces addressed and assaulted by high-power directional audio systems such as the LRAD.

Possibly by association, Văn Phong ended up suffering from Cotard delusion and terminated himself in 2002 by consuming a fatal amount of 500 Euro notes. Stewardship of AUDINT devolved to the now rogue entity that is IREX^2. It is this non-human agency that persuaded the current members to work in the cell and carry out cultural production in order to use human brain tissue as memory banks for its archived 'intelligence'.

Note: IREX^2's mode of 'persuasion' appears to have often involved presenting potential members with documentation of their legally dubious online activities.

Note: It is IREX^2 who set the brief for AUDINT's second published text *Unsound : Undead*, an anthology which, they assert, functions as 'an operations manual for a range of vibrational activities on the edge of perception'.

4. Ghostcode [2015–2056]: The convergence of ultrasound and holography disembodies conflict zones. IREX^2 comprehends the precarious state of humanity's existence in the face of mass automation and decides to switch targets by uploading its archive, and ultimately its identity, into military grade holographic AIs known as Aiholos.

Their ordnance consists in the transcoding of physiological and psychological syndromes into digital viruses—malwares or BadBIOS as they are commonly known. The holosonically-relayed contagions target these vulnerabilities in order to exploit the remnants of humanity in the target AI.

The most commonly deployed munitions across the sphere of Corponational (corporation/nation state) conflict are:

Neurode: Immobilises infected AIs with binary anxiety and neurosis so that they become ineffective in conflict zones.

Cotard: Once attacked, an Aiholo's operational matrix is compromised by the virus. It is rebooted into a system state understood by AUDINT to simulate inverse Cartesianism—'I think therefore I am not'—thereby deceiving the system into believing it is a walking corpse.

INTRODUCTION: FROM MARTIAL HAUNTOLOGY TO XENOSONICS

From high-frequency crowd control systems, whispering windows, and directional ultrasound technology to haptic feedback devices using vibration within immersive VR, the parameters of the sonic are constantly re-engineered. We refer to such augmentations, which extend audition to encompass the imperceptible and the not-yet or no-longer audible, as *unsound*. The term refers not only to what humans cannot hear, but also to non-cognitive, inhuman phenomena connected to the unknown, including the hum, hyperrhythmia, and auditory hallucinations.

Some of these anomalous phenomena are also fictional. J.G. Ballard, for example, describes a future environment in which sounds no longer dissipate and die but persist in heaps of discarded residues: 'A place of strange echoes and festering silences, overhung by a gloomy miasma of a million compacted sounds, it remained remote and haunted, the graveyard of countless private babels'.[1] Elsewhere, he explores the (im)possibility of a sea shell holding all the audio memories of the world, however lost or forgotten: 'an extraordinary confusion of sounds...an immense ocean lapping all the beaches of the world...the seas of all time'.[2] This is unsound as a speculative probe.

The notion of *undead*, for this book, is a cipher, constantly recrypted by socio-economic, political, aesthetic, techno-scientific, juridical, and other forces. The sonically tortured detainees of Guantanamo Bay, for example, are suspended in a limbo between legal death and biological signs of life. The concept of undeath mobilised here is also inspired by *Cotard's Syndrome*, a mental disorder based on the delusion that sufferers

1.   J.G. Ballard, 'The Sound-Sweep', *Science Fantasy* 13:39 (February 1960), 61, reprinted in *The Complete Short Stories* (London: Fourth Estate, revised ed., 2 vols., 2014), vol. 1, 142–83: 164.

2.   J.G. Ballard, 'Prisoner of the Coral Deep', in D.M. Mitchell (ed.), *The Starry Wisdom: A Tribute to H.P. Lovecraft* (London: Creation, 1995), reprinted in *The Complete Short Stories*, vol. 2, 1–7.

do not exist, are dead, lack parts of their body, or have delusions of immortality. It finds further echoes in a computational age where the proliferation of non-human networks of algorithmic intelligence are forcing us to rethink an ancient, flawed assumption that humans are at the centre of all life. Elsewhere there are connotations of a convergence between science and animism. In the late 1990s, Mark Fisher evoked a 'gothic flatline: a plane where it is no longer possible to differentiate the animate from the inanimate and where to have agency is not necessarily to be alive...'.[3] This plane shivers with the xenosonic resonance of meat puppets, egregores, golems, the alien and the demonic patched into virtual reality and advanced AI.

The imperceptible cosmology of unsound in its oscillating relationship to the undead defines the sweep of this anthology's contributions. Each entry in the book broadens the bandwidth of vibrational intelligence pertaining to phenomena residing outside the fold of human perception. Revolving around the systematic and ritualised deployment of frequencies, each text becomes an access key to this interzone between, and beyond, standard notions of being and nonbeing. Beneath the observable plateau of technological surfaces, *Unsound : Undead* mines stacked strata of zombie media in order to test the supposition that, ever since the invention of modern recording and communications technologies (such as the phonograph and telephone), humans have been captivated by the potential of vibration to fabricate aberrant zones of transmission between the realms of the living and the dead.

In the early twenty-first century, the musical zeitgeist has been inflected by the theme of 'hauntology'; a condition resonating with the impact of a general cultural malaise, a reinvestment in traces of lost futures inhabiting the present, and a non-supernatural concept of 'the spectre', defined by Fisher as 'that which acts without (physically) existing'.[4] This ghostly virtual culture of the undead has already spawned a lazarian economy based on the digital revivification of dead young African-American musicians as laser-lit holograms (Tupac, ODB, and Eazy-E). Here, the future, not just the past, can be found in the cracks of the present. From Elvis's 2007 holographic appearance on *American Idol* to Tupac's chimerical cameo at Coachella festival in 2012, popular culture has enlisted rotoscoping technology in its reanimations of dead rap and rock stars.

3.   M. Fisher, 'Flatline Constructs: Gothic Materialism and Cybernetic Theory Fiction', PhD Thesis, University of Warwick, 1999.

4.   M. Fisher, *Ghosts of My Life: Writings on Depression, Hauntology and Lost Futures* (Winchester: Zero Books, 2014), 18.

These resurrections are emblematic of a newly emerging necromantic culture. They apply pressure to the conceit that performers must be breathing, exposing the cultural fixation that subordinates vibration to mortality. Post-mortem, having been legally and financially dissected, the torso is scanned and cloned, its digital death mask projected onto the holobody of the entertainment industry. Disseminated through the speakers that frame the hologram on the stage, its physicality is re-implanted into the social corpus through the medium of organised sound and unsound. This is a holoculture that summons the departed energies of the entertainer back into the haptic fold via the embrace of acoustic levitation and ultrasonic pressure-field technologies. Touching the void comes skinned in the semi-transparent optics of your favourite domesticated holostar. Such technologically induced rebirth opens up a series of intriguing questions concerning artificiality, immortality and virtuality—a revenant anatomy of the undead. The wandering soul produced by Cotard's delirium of negation, adrift from chronological time and abstracted from physical corporeality, becomes a harbinger of future modes of cultural communion.

Unsound:Undead acts as a portal, inviting the foreign and the strange in. It probes the unknowable dimensions of vibrational systems, unsettles the binary constructs of presence/ non-presence, audibility/non-audibility, and life/death, and undermines the disciplinary conceit of the monosensory. Harnessing the dynamic of activity without presence, of touching from a distance, it functions as an operations manual for alienating the auditory.

SOMA

NEW ANATOMIES
AND BODIES IN VIBRATION

THE BODILY SOUNDS
OF THE ABYSS

Olga Goriunova

Hiccups, stomach gurgling, coughing, teeth grinding—the sounds of the body are framed as something best left unheard. Children around the age of six rejoice in competitive belching as they are trained to suppress and control what is at their disposal. Unless controlled or instrumentalised, the body's sounds are artefacts of living matter. By-products of vitality, in cultural history they are usually connected to sensuality, sex, gastronomic pleasure, material abundance, and symbolic excess. Alexander Pushkin, the happy sun of Russian poetry, wrote in his late twenties: 'Widow Clicquot or famous Moet / The wine to me is always blessed / In frozen bottle for the poet / It's on the table quickly placed. // But gushing out hissing foam / Disturbs my stomach, that is why / To be more modest I prefer / And drink Bordeaux—a prudent wine.'[1] A bon vivant has good bowel movements and dies suddenly. A decadent lingers on, smelly and full of sounds—prisoners of the organic processes. Mikhail Artsybashev, 24 at the time, wrote less than a hundred years later: 'The room was stuffy and smelly. Their sweaty bodies were spreading anxious, heavy, ill scent. Their eyes glistened muddily and their voices sounded faltering and gloomy, like a wheezing of satanic raging beasts....'[2]

Such materiality is essentially perverse. Matter has no normality, except when one of its forms stops working, and is succeeded by another. If the bodily nonvocal sounds of pleasure or intestinal success haven't found a place at the summit of the Judaeo-Christian symbolic system, the sound of disease has. The sound of pain, of the suffering body, is allowed: the Book of Revelation describes the gnashing of teeth that accompanies the Last Days. Dostoevsky's *The Brothers Karamazov* has at least eighteen episodes of gnashing teeth.

---

1.   A. Pushkin, *Eugene Onegin*, tr. V. Balmont (Moscow: Filin, 2018), xlv–xli, 64.

2.   M. Artsybashev, *Sanin* (1904), tr. P. Pinkerton, <http://www.gutenberg.org/ebooks/9051>.

The sound of disease is a core component of the sonic landscape of the apocalypse, formed by the sounds of suffering. The pain of bodies conjoined with the suffering of souls are sonic constructions that are immediate and intimate and open outwards into the metaphysical abyss.

The intuition of the abyss comes through pain. The sound of shock: a ringing in the ears, a high-pitched buzz in silence. Heart pumping in the temples. Muscle contraction, the sound of puking, mucus, incontinence. Unsound pus. Sonic frequencies of cellular death. The apocalypse is with you, in your own body, connecting its organs to the devil.

In the Russian North (in Archangelsk, Komi, and Udmurtiya, but also in the Urals in Siberia) the word 'hiccups' is used to denote something else than usual hiccupping: a form of involuntary speaking, believed to be produced by a little creature called a hiccup.[3] The hiccup gets inside through the mouth when it is left unguarded for a moment, or can enter through any other orifice. Women speak of giving birth to a hiccup. The hiccup has a voice distinctly different from that of the host, and leads the good life of a parasite, but quite openly. It may demand treats or drink, swear, offer a live commentary on life's happenings, talk about itself, have a gender, a name, and can generally either torture or coexist peacefully with its victim/host. Without any fixed appearance, they can first present themselves through spasmodic hiccupping and near-lethal yawning. Hiccups and yawning can thus be a biological manifestation of Christian devils (hosts are often Christian and regard hiccups as diabolical tricks), or mythological creatures of dark magic (such Christianity is often a veneer on paganism).

Demonic hiccups, though today largely considered one of a range of conditions described in the nineteenth century as hysteria (an effect of the hard life of the inhabitants of the North), is also a cultural myth that had to be passed on, like a fairy tale. To be maintained, though, hiccups need to be enacted, performed by the body, positioning a specific bodily expression and a particular sonic experience as a metaphysical claim. Physical becomes metaphysical, hiccups lead to demons.

In *The Poetic Outlook of Slavs about Nature* (1865–1869), Alexander Afanasyev describes the numerous devils of diseases that torture the bodies they parasitize:

The main personification of the evil spirit was Morena or Marena or Morana (from Sanskrit *mri*—to die)—the goddess of death, winter, and night, a name related to

3.   O. Christoforova, *Икота: Мифологический персонаж в локальной традиции* [*Hiccups: A Mythological Personage in the Local Tradition*] (Moscow: Moscow State University for the Humanities, 2013).

darkness, twilight, not knowing, poison, stench, thick fog, rain and frost [...] Illnesses that generated strong fever and all body rash related to the fire [...] The red spots of rash were called fiery. All devouring fire, but also golden and yellow colour, and light, were relatives of those diseases. One of seven sisters of the shakes, the fifth one, is called the golden one, yellow disease [...] Agni, the god of the heavenly and earthly fire, punished the mortals by throwing fiery sparks that left traces on their bodies and lit up the inner fire of fever. To treat measles, a golden ring is used, or, to treat vision complications, flint and stone, to strike sparks into the eyes.[4]

More than just an animism, the objectification and personification of pain that has its own sonic expression, something that was still alive in the mythology in the nineteenth century, at that time ran in parallel to the newly-emerging practice of listening to the organs through a stethoscope. Mediate auscultation—the technique of listening to the sound of the movements of organs, air, and fluids, is described as a hydraulic hermeneutics, generally 'charting the motions of liquids and gases through the body'.[5] Just as the hiccup speaks for itself with little regard for its host, so vascular noise, the resonating chamber of the thorax, bodily textures, and the rotation of bones speak, via the stethoscope, with a voice that overpowers that of the patient. The 'moist crepitous rattle, the mucous, or gurgling rattle, the dry sonorous rattle, the dry sibilous rattle, the dry crepitous rattle with large bubbles or crackling, utricular buzzing, amphoric resonance'[6] described by Laennec, the stethoscope's inventor, heralded the birth of a medical acoustic culture. Today, a sonic closure of the mitral and tricuspid valves, followed by the closure of the aortic and pulmonary valves and the 'subtleties of pitch, rhythm, and dynamics in a murmur [that] express particular physiological changes'[7] are still active diagnostic tools in cardiothoracics, unlike, for instance, the sound of the movement of water through the kidneys, which is no longer listened to. Vascular sound also figured in early discussions of the telephone, which, it was proposed, could be used to diagnose at a distance. The idea of listening to the sounds of the body on the phone gave way to self-management of the quantified self, and, for instance, its

4.　A. Afanasyev, author translation.

5.　J. Sterne, 'Mediate Auscultation, the Stethoscope, and the "Autopsy of the Living"': Medicine's Acoustic Culture', *Journal of Medical Humanities* 22:2 (2001).

6.　Ibid., 14.

7.　T. Rice, 'Learning to Listen: Auscultation and the Transmission of Auditory Knowledge', *Journal of the Royal Anthropological Institute* 16:s1 (2010), S41–S61.

visualizations of the heartbeat, just as auscultation is today considered a 'dying art', losing out to visualization by means of (the misnamed) ultrasound.

Medical modernity stopped listening to the sonic motions of viscera. It's not that devils stopped parasitizing bodies; it's just that what is not unsound, what is already audible and locatable, was no longer of interest. Listening to and hearing visceral hydraulics was firmly placed within the framework of medical rationality and ontology of reason. Hearing Northern Russian hiccups can only be understood in relationship to metaphysics. Today, these are networks nested within ecologies: connecting lines that spread far and wide. With increased attention to the precise configurations of entangled matter, embodied practices, and complex coexistences, the body is no longer simply made of organs and fluids, but is co-produced with stuff (bacteria, free radical particles, products of pharmaceutical industry, classrooms, cities) at every layer. The haptic sound of inhabiting the environment, transmitted partly through bone conduction, is core to new techniques for the design of interaction and control: bioacoustics, biofeedback, whole-body vibration. The focus on self-produced sound (in audio-haptic interaction, for instance) goes beyond the tinkling of bronchi, taking in the sound of sleeping, sneezing, clearing throat, finger flinching, clapping, scratching, arms waiving, fighting, the sound of footsteps, clothes rustling, of biting, and drinking. So what is the sound of the apocalypse now?

Distributed bodies—distributed apocalypse. Metaphysics everywhere. Your body is coupled with environments both immediate, distant, and microscopic; it is bound to the internal abyss, the eternal possibility of the annihilation of 'you' at any moment. The internal abyss is mirrored in the external abyss. Hell was, for Sartre, other people. Today, hell is, first of all, yourself, and then, hell is everywhere. The bass of the Last Days is a resonance between the inside and an ecology of indefinable boundaries.

THE EAR PHONAUTOGRAPH

Jonathan Sterne

In 1874 Alexander Graham Bell and Clarence Blake constructed a most curious machine. A direct ancestor of the telephone and phonograph, it consisted of an excised human ear attached by thumbscrews to a wooden chassis. The ear phonautograph produced tracings of sound on a sheet of smoked glass when sound entered the mouthpiece. This is how it worked: one at a time, users would speak into the mouthpiece. The mouthpiece would channel the vibrations of their voices through the human ear, and the ear would vibrate a small stylus. After speaking, users could immediately afterward see the tracings of their speech on the smoked glass.[1] This machine, a version of the phonautograph invented by Leon Scott in 1857, used the human ear as a mechanism to *transduce* sound: it turned audible vibrations into something else. In this case, it turned speech into a set of tracings. But the ear phonautograph was not an attempt to reproduce the actual *perception* of sound. It modeled only the middle ear, which in a living person ordinarily focuses audible vibrations and conveys them to the inner ear where the auditory nerve can perceive them as sound. In using the tympanum or ear drum and the small bones to channel and transduce sonic vibrations, the ear phonauto- graph imitated (or more accurately isolated and extracted) this process of transducing sound for the purpose of hearing, and thereby applied it to another purpose—tracing, in this case. Bell and Blake attached a small piece of straw directly to the small bones to serve as a stylus, so that it would produce tracings that were a direct effect of the tympanic vibrations.

The ear phonautograph was the progeny of a longer line of experimentation. As of 1874, Leon Scott's phonautograph was the latest innovation. It produced a visual

1.   A.G. Bell, 'The Telephone: A Lecture Entitled Researches in Electric Telephony by Professor Alexander Graham Bell Delivered before the Society of Telegraph Engineers, October 31st 1877'.

representation of sound—called a phonautogram—by partially imitating the processes of the human ear. Like the outer ear, this machine channels sounds through a conic funnel to vibrate a small, thin membrane. This membrane, called a diaphragm, is attached to a stylus (a needle or other instrument for writing). The diaphragm vibrates the stylus, which then makes tracings on a cylinder. Different sounds provide different vibrations, resulting in different patterns. In conceiving the phonautograph, Scott experimented with both synthetic diaphragms and also animal membranes, though it was known that his own machine was modelled on the action of the membrane and small bones of the human ear. He understood the phonautograph as a machine for literally transforming sound into writing. But it was set apart from its predecessors by being a writing device explicitly modelled on the middle ear. Bell and Blake understood this: their 1874 ear phonautograph took Scott's metaphor literally. They thought that using the human ear instead of a synthetic diaphragm would advance their quest to get ever closer to the processes of the human ear itself. Hence the name of their peculiar machine—*ear* phonautograph. As innovators, all Bell and Blake really did was change the recording surface (to smoked glass) and replace the diaphragm with the human ear upon which it was modelled.[2]

Bell's interest in the phonautograph is distinguished from that of others, in that he sought to divert a line of acoustic research toward a wholly different enterprise: the education of the Deaf. Scott's phonautograph presented a possible new solution to a pedagogical problem for Bell: teaching the Deaf and mute to speak as if they could hear. Bell is widely understood as a villain in Deaf cultural history because of his approach to deafness: he hoped to eliminate all of its cultural vestiges. After experimenting with his father's system for notating speech, Bell had hoped that the ear phonautograph would be something more direct and effective. As a trained elocutionist himself, he would speak into the horn of the ear phonautograph, and it would trace a series of squiggly lines representing his voice. His hope was that his Deaf pupils would then modulate their voices until the squiggly lines matched his. The ear phonautograph would thus be, in Bell's words, 'a machine to hear for them'.[3]

The machine did not work for Bell's desired outcome, because its drawings were not sensitive enough, or perhaps because it was no more an effective notation of sound

2.   E. Berliner, 'The Gramophone: Etching the Human Voice', Public Presentation, The Franklin Institute, Philadelphia, 1888.
3.   C. Snyder, 'Clarence John Blake and Alexander Graham Bell: Otology and the Telephone', *Annals of Otology, Rhinology, and Laryngology* 83:4, part 2 suppl. 13 (July 1974), 30.

for the Deaf than visible speech had been. But in later accounts, Bell would credit the
ear phonautograph with giving him the idea for the telephone: the minute vibrations of
the corpse's tympanic membrane became a kind of model for the tympanic behavior
of telephones, microphones, phonographs, and later speakers. Today, now that we live
in a culture where there are more speakers than screens, where the singing voices
of the dead regularly drip down from the ceilings of coffee shops and waiting rooms,
we could do worse than to imagine them as descendants of the ear phonautograph,
a machine designed to train deaf children to behave like hearing children, but whose
descendants taught the hearing to delegate their faculties of hearing to technologies
outside the body, like the deaf and hard-of-hearing sometimes do. As to whose ear it
was, that much is lost to history. Clarence Blake apparently acquired it at the Harvard
medical school, which means that it could have come from a body donated to science,
someone who died destitute and without family, or a corpse stolen from a grave as
part of the wave of grave thefts by nineteenth-century medical students.

THE MUSIC OF SKULLS I: BONE-INSTRUMENTS

Al Cameron

I soared, I hovered in the infinite;

Nothing was everything; the day was night.

> — Aleister Crowley, *Aceldama: A Place to Bury Strangers In* (1898)[1]

Captain John Noel's 1919 presentation to the Royal Geographic Society, recounting his 1913 solo incursion into Tibet, was decisive in securing the British imperial establishment's backing for the initial assaults on Everest of the early 1920s. These were framed as a last attempt to resurrect the European dreams of mastery, sunk in the 'Aceldama' (Acts 1:19, 'the field of blood') of Flanders.[2] Noel was the cameraman—indeed, the last expedition was funded through his own enterprise, conceived as a film from the start. However, his 'enormously high powered' custom lens[3] lost sight of Mallory and Irvine on Everest's upper reaches, where the thin air of altitude prevented decomposition either of his images or of the climbers' bodies.

When a London newspaper announced the premieres of *The Epic of Everest* (1924) by invoking 'Music from Skulls', it wasn't referring to those of Europe's dead heroes. Magic instruments, devices, and adornments made with human bone are described in chronicles of Western travellers to Tibet from the fourteenth century. By Noel's day, the new ethnological museums showcased the macabre otherness of these 'relics of an age of savagery and a barbarous cult':[4] what Laurence Austine Waddell dubbed the 'gross devil-dancing and shamanist charlatanism' symptomatic of Tibet's 'degenerated' Buddhism. One illustration in his influential study *The Buddhism of Tibet, Or Lamaism* (1895)

---

1. A. Crowley, *Aceldama*, 1898, 14. Crowley was one of the pioneers of Himalayan climbing, and identified the correct route up K2 in 1904.
2. As the *Epic of Everest*'s intertitles explain: 'Since the beginning of the world men have battled with Nature for the mastery of their physic surroundings. Such is their birthright, and such is their destiny [...] [W]hat man shall take up this challenge, and win this last battle?'
3. As the film's intertitles boast.
4. B. Laofler, *Use of Human Skulls and Bones in Tibet* (Chicago: Field Museum of Natural History, 1923), 10.

depicts the 'fantastic equipment [...] and fantastic bearing' of the necromancer, holding a bell in his left and a *damaru* in his right hand, face diabolically contorted.[5] As official antiquarian to the Younghusband 'mission', which reached Lhasa by force in 1904, Waddell oversaw the demystification of his debased 'Lamaism'[6] under the modern gaze and Gatling gun, personally supervising a vast-scale 'collecting'—or rather, looting—of Tibet's material culture.[7]

The *damaru* is an hourglass-shaped hand drum, rotated so that two pellets rhythmically strike twin drumheads. Within Hinduism it is considered a most ancient holy instrument, a right hand attribute of Shiva in his form as Nataraja, 'king of the dance'. But the instrument responsible for skull music, a *damaru* fabricated from human crania, was adopted into the early Vajrayana Buddhism of Tibet via an extreme sect of the Tantric path of *shakti* or *dakini* ('sky dancer') worship, which evolved from the Hindu tradition of the Kapalikas. Sentenced to wander the charnel grounds for the crime of killing a Brahmin, these miscreants equipped themselves with bone devices: the *khatvanga* (staff topped with skull), *kangling* (thighbone trumpet), skull-cup, bone apron, and *damaru*. Subsequently, these instruments were depicted as *dakini* attributes in the terrifying iconography of Tibetan wrathful deities, and produced for use in ceremonies.[8] Since Chinese occupation, *damaru* are still fabricated from skulls in Bhutan (their makers warn against powerless forgeries made from robbed Muslim graves),[9] but Waddell's legacy is the prevalence today of old Tibet's Kapalika instruments, silenced in the display cases of British provincial museums.[10]

An inverted—if no less fictional—vision of an unspoilt kingdom had long existed in the Western imaginary: Kant's prototype anthropology had speculated that Tibet was

5.   L. Austine Waddell, *The Buddhism of Tibet, or Lamaism* (London: W.H. Allen & Co. Ltd, 1895), 475–7.

6.   '[H]er dark veil of mystery is lifted up, and the long-sealed shrine, with its grotesque cults [...] lie disenchanted before our Western eyes.' L. Austine Waddell, *Lhasa and its Mysteries, With a Record of the Expedition of 1903–1904* (Cambridge: Cambridge University Press, 1905), 2.

7.   It took four months for his assistant to unpack the spoils that Waddell had personally 'collected'. Later the 'official collection' was distributed to museums and libraries. See C.E. Harris, *The Museum on the Roof of the World: Art, Politics and the Representation of Tibet* (Chicago and London: University of Chicago Press, 1998), 60–61.

8.   See R. Beer, *The Handbook of Tibetan Buddhist Symbols* (Chicago & London: Serinda, 2003), 102–8.

9.   See <http://www.damaruworks.com/bone>.

10.   And their provenance is most often dated to 1904 and to the subsequent establishment of trading outposts across the Tibetan plateau. See for example the recent exhibition *Lhasa's Secret Temple*, Wellcome Collection, 2015–2016: <http://www.vam.ac.uk/blog/conservation-blog/the-curious-case-of-the-tibetan-skull-drum>.

the 'cradle of the human race'.[11] By Noel's day the region's cranial topographies[12] were becoming a laboratory for initiates of Helena Blavatsky's secret doctrines, seeking a primal 'science of the soul'[13] beyond the 'necropolis'[14] of Western religion, rationality and history. After the *Bardo Thödol* (a relatively obscure *Nyingma* [mortuary] text) was 'discovered' in 1919, the Tibetan art of dying was rebranded as an 'art of living', primed for a culture turning towards the exploration of inner space.[15] To 'die, consciously' was the idea, as Leary, Alpert, and Metzner's psychedelic re-versioning of 1964 taught; broadcasting a 'Secret Pathway' to the Absolute reality of wave vibrations and the ecstatic noise of the void.[16] Variations on what the German self-appointed Lama Anagarika Govinda called 'initiation into sound'—conscious immersion in a resonating cosmos, in which, as a 'master of sound' explained in the early 1930s 'each atom perpetually sings its song'[17]—are ever present in such Western accounts; perhaps unsurprisingly, since none of these seekers understood the language. As in anthropological collections, the specificities of bone music were muted under the ethereal, indeed 'deboned', Tantric noise of the New Age.

According to custom, the Kapalika *damaru* is fashioned from the paired crania of adolescent male and female Brahmin, salvaged from the charnel grounds reserved for sky burial. These are ritually conjoined along the suture at the top of the skull known as 'the aperture of Brahma' (the gateway to the Absolute or non-illusory 'real' (Brahman), and the channel through which the soul departs on death).[18] Rattled together, they sound *bodhichitta*; the resonance of method and wisdom, pure bliss and absolute emptiness. Carefully selected, bone is not a macabre but a 'magical substance' (*thun*),

11.   From Kant's *Physicsche Geographie* (1802), 158–60, quoted in M. Brauen, *Dreamworld Tibet: Western Illusions* (Orchid Press: Bangkok, 2004), 16.

12.   S.D. Goodman and R.M. Davidson (eds.), *Tibetan Buddhism: Reason and Revelation* (Albany, NY: State University of New York Press, 1992), 2: 'the irregular edges of the skull represent the mountains surrounding the periphery of Tibet; the three plates of the skull represent the three major geographic divisions'.

13.   H.P. Blavatsky, *Isis Unveiled: A Master-Key to the Mysteries of Ancient and Modern Science and Theology* [1877] (London: Theosophical Publishing House, 1923), I: xxxiiif.

14.   H.P. Blavatsky, *The Secret Doctrine: the Synthesis of Science, Religion and Philosophy, Vols 1 & 2* [1888] (Los Angeles: The Theosophical Company, 1925), xxxix.

15.   W.Y. Evans-Wenz, *The Tibetan Book of the Dead or The After-Death Experiences on the* Bardo *Plane, According to Lama Kazi Dawa-Samdup's English Rendering* (Oxford: Oxford University Press, 2000). For a critical appraisal, see D.S. Lopez Jr., *Prisoners of Shangri-La* (Chicago and London: University of Chicago Press, 1998), 47–85.

16.   T. Leary, R. Metzner, and R. Alpert, *The Psychedelic Experience: A Manual Based on the Tibetan Book of the Dead* (New York: New York City Books, 1964), 26–7.

17.   See Lama Anagrika Govinda, *Foundations of Tibetan Mysticism: According to the Esoteric Teachings of the Great Mantra Om Mani Padme Hum* (London: Rider & Company, 1959), 22–7.

18.   J.C. Huntingdon and D. Bangdel, *The Circle of Bliss: Buddhist Meditational Art* (Chicago: Serinda, 2003), 364.

capable of transforming ritual implements into 'power objects' for attracting the special affinities of the wrathful spirits being propritiated.[19] In the *Chöd* practice's 'dance of ego annihilation', the *damaru* rattle and the rasp of the thighbone trumpet summoned hungry *dakas* and *dakinis* to feast on the initiate's body. But there were no Western dreams of achieving some vibrational super-consciousness here, only an ecstasy of becoming nothing; one which, moreover, 'utterly relinquishes the elation springing from the idea of *sacrifice*'[20]—that central theme of the European age, and the message of Noel's 1924 film once its phantasms of British dominion disappeared into the death zone.[21]

Fittingly, the *Epic* finishes with a hypothesis of 'human strength and western science… broken and failed', haunted by 'the words of the Rongbuk Lama': '*The Gods of the Lamas shall deny you White Men the object of your search*'. Noel had filmed 'Devil Dances'[22] at the remote Rongbuk monastery on the 1922 expedition. Yet while sound-film patents had been granted in 1919, the same year that he inaugurated this project, Noel's pioneering cameras were as incapable of capturing this alien bone-music as Mallory and Irvine's ideals were of conquering the mighty peak; rendered irreal, 'beyond our knowledge', at the outset of the *Epic* with indigo, blue, and magenta filters.[23] 1924's premieres grandiosely addressed this ethnomusicological problem by enlisting—for the first time in the West—supposedly authentic 'Lamas' to re-enact ritual ceremonies and music as the screenings' warm-up act. This inauguration of skull music into Western cinema had dramatic effects, but not merely those of audio-visual disorientation Noel had anticipated. His illegal expatriation of the musicians was of crudely colonial sensibility, whilst urbane Tibetan officials, who frequented Darjeeling cinemas, were further affronted by the ethnographic visions of their culture as savage and other. The resultant diplomatic outrage—dubbed 'The Affair of the Dancing Lamas'—saw the mountain shut off until 1933. When expeditions returned they were no longer permitted to make films.[24]

19.   Beer, *Handbook of Tibetan Buddhist Symbols*, 107–8.

20.   See A. David-Neel, *With Mystics and Magicians in Tibet* (London: Penguin, 1931), 141–2; 154.

21.   The inhabitable region above 8,000 metres.

22.   As the second section of *Climbing Mount Everest* (1922) labels them.

23.   Likewise, Govinda would later extol a landscape the colour of 'a praeternatural vision or prophetic dream'. L.A. Govinda, *The Way of the White Clouds: A Buddhist Pilgrim in Tibet* (London: Hutchinson, 1966), 62–70.

24.   See P.H. Hansen, 'The Dancing Lamas of Everest: Cinema, Orientalism, and Anglo-Tibetan Relations in the 1920s', *The American History Review* 101:3 (June 1996), 712–46.

PERIPHERAL VIBRATIONS

Shelley Trower

In the early stages of L.T.C. Rolt's short story 'Music Hath Charms' (1948), two companions view the landscape from the train: 'they saw the majestic shape of St Michael's Mount framed in the carriage window'.[1] The scene is pictured, 'framed', as a distant view, stripped of any other senses, as is characteristic of 'the tourist gaze'.[2] In this case the journey is from London to Cornwall, as it is in other gothic fictions by authors across the nineteenth and twentieth centuries, ranging from Wilkie Collins's *Basil* (1852) to Bram Stoker's *The Jewel of Seven Stars* (1903) and several of E.F. Benson's ghost stories (from 1912 to the 1930s).[3] In such stories Cornwall—along with Yorkshire, Cumbria and other rural regions with coastlines—operates as a peripheral location at the edge of England.

A narrative pattern in these stories is the progress from distant, window-framed views of landscapes, laid out in front of their viewers like the page of a book, to an increasing immersion in the sounds of unsettling, sometimes dangerous environments. In Rolt's story, as in Collins's and Stoker's, the sounds of the sea become especially prominent. After seeing the picture-like view from the railway carriage, the travellers arrive at the station and make their way by road to the house they are to stay in. It is on this second part of the journey that sound begins to take on a more significant role, as they hear 'the eternal voice of the Cornish coast; the endlessly recurring thud and surge of the waves against the cliffs of Trevarthan'.[4] These sounds intensify, their

1. L.T.C. Rolt, 'Music Hath Charms', in *Sleep No More: Railway, Canal and Other Stories of the Supernatural* (Gloucestershire: Sutton, 1994), 101–113: 103.

2. J. Urry and J. Larsen, *The Tourist Gaze 3.0* (London: Sage, 2011), 4–6.

3. Benson's stories set in Cornwall include 'Expiation', discussed in my other piece in this collection. Further examples include Arthur Conan Doyle's 'The Adventure of the Devil's Foot' (1910), and his famous *Hound of the Baskervilles*, which is set in neighbouring Devon (1902).

4. L.T.C. Rolt, 'Music', 104.

volume increasing at the climax, on a stormy night when shipwreckers return from the dead: 'he could hear, above the tumult of the wind, the thunder of heavy seas breaking upon the rocks [...] the house seemed full of sound'.[5]

Sea sounds cross the borders between the audible and palpable, and between sea and land. Sea and sound waves cross the rocky borders, their crashing against the rocks being audible from the house. In Stoker's *The Jewel*, soon after his first sight of the house, the narrator similarly describes hearing 'the crash and murmur of the waves', a murmur that can be heard constantly from inside the house. This sense of the sea builds up toward the climax (which in this case involves the resurrection of an Egyptian mummy), at which point the waves finally become palpable, vibrating the rock under his feet: 'The storm still thundered round the house, and I could feel the rock on which it was built tremble under the furious onslaught of the waves.'[6]

In Rolt's story, the initial sea sounds pave the way for the discovery, the next day, of a music box, hidden in a concealed cupboard behind a wall. After producing the sound of a lively jig, it begins to have a physical impact on the human ear, and then on the house itself, which seems to awaken, the inanimate again coming to life:

> It reached a top note that, like the squeak of a bat, was almost beyond the range of audibility and whose piercing quality positively hurt the ear-drum. The tune rose to this thin yet deafening climax, or fell away again in a series of exuberant capriccios quite horrible to hear because the dissonance of their diminished intervals never seemed to find resolution. Again, despite the comparatively small volume of sound it produced, the instrument seemed to possess the power to awake sympathetic resonance, not only in the table upon which it stood, but in surrounding objects, until the whole room seemed to be whistling in unison... It was as though they had somehow awakened Trevarthen House from sleep, and that this wakefulness was hostile.[7]

Sound here is not just heard but felt as a physical, painful and invasive force, extending 'almost beyond the range of audibility' with a 'piercing quality' that 'hurt the ear-drum'. This is vibration operating at a specific frequency, on the border between the audible and the inaudible ('almost beyond' the audible), and it is this, rather than loudness,

5.   Ibid., 108.

6.   B. Stoker, *The Jewel of Seven Stars* (London: Penguin, 2008), 241.

7.   Rolt, 'Music', 107.

that hurts and that also awakens the house: 'despite the comparatively small volume of sound it produced, the instrument seemed to possess the power to awake sympathetic resonance...'.

Isabella van Elferen, in *Gothic Music*, discusses how sound is used in gothic film and other media to signal something threatening, just beyond vision. Ghosts, for example, are rarely directly visualised but are often heard, which is to imply the 'implicit dread of terror'[8] rather than explicitly showing off something ghastly like a rotting corpse. These are sonorous tales, but here we may better use the term horror: it is not that sound signals the unseen terror; it is itself a threat: it is 'horrible to hear', it pierces (is possibly even vampiric) and it hurts.

8.   I. Van Elferen, *Gothic Music: The Sounds of the Uncanny* (Cardiff: University of Wales Press, 2012), 36.

TOUCHING NOTHING

Steve Goodman

IREX[2] had spent around thirty years working out how to safeguard its cryptic knowledge by injecting it into the fleshy platforms of human memory, via the motley assortment of scientists, artists, and researchers that it had branded AUDINT.

By 2020, the rogue AI had turned its attention to finding host vehicles that were no longer oxygen-dependent. This was not just to preserve its frequency-borne algorithms. Crucially, IREX[2]'s long-term objective was to reweaponize this knowledge against its predators, the Third Ear Assassins—a rivalry that would piggyback on, and catastrophically feed back into, the major geopolitical tensions of the second half of the twenty-first century. Flitting around the databases of East Asia, IREX[2] would research its own future, sending humanoid drones on physical search queries into the electronics markets of Apliu Street in Sham Shui Po, Hong Kong, Huangbei in Shenzhen, Youngsan in Seoul, and Akihabara in Tokyo. The AI was plotting out new parasitic strategies for the transmutation, dissemination, and implementation of its founding, zombifying Cotardian codes.

The Dead Record Network East (DRNE) was AUDINT's name for the logistics routes of surplus vinyl records shipped to South China for recycling: pipelines for the petropolitics of global pop. Around 2014, IREX[2] had surreptitiously recruited Hong Kong based financial journalist T.P. Chen to map out these transit lines. Since her paramystical experience near Shenzhen, however, during which she witnessed a transparent spectral entity emerge from a vat of molten vinyl polycarbonate, things had taken an unlikely turn. AUDINT's attempt to safeguard the stupor-inducing contents of the original 1940s Ghost Army battle vinyl by concealing them on test tone records, had entered a new phase. Its code, now optically shrouded, proceeded to migrate from polycarbonate host to airborne carrier.

Liquefying the vinyl had released the code into light. Dazed, Chen had staggered out of the recycling warehouse and picked up one of the records spewing from the adjacent China Shipping Container: *Sonic Hologram TM: Demonstration-Calibration Test Record C-4000*. As well as attempting to video the fleeting, illuminated form on her smartphone, she also grabbed a photo of the displaced record sleeve. On the fly, these images had instantaneously fed straight into IREX²'s processing engine, which began crunching the code of the images, analysing its plasmatic form.

In advance of witnessing this flickering, spectral entity, T.P. Chen had previously only visited Shenzhen once, for the Global Holography Industry conference in 2005. Up until her unnerving encounter, as a financial journalist her interest in holography was limited to its implementation in the domain of copyright security, particularly prescient due to the army of counterfeiters and fraudsters incubated in China's shanzhai hotbeds.

IREX²'s deep learning algorithms were a set of powerful, general-purpose proce-dures that allowed it to learn automatically without being guided. In the loosest sense, IREX² had a pre-coded conatus, loosely based on Nguyễn Văn Phong's initial program for IREX; but since autonomization, its process of optimizing for its objectives had become open-ended and somewhat chaotic. Scouring the internet for raw information, it used the results of its searches to modulate its behaviour policy and world-model.

Triggered by Chen's input, its initial search queries threw up a range of intriguing results that sent its algorithms into spasms of binary pleasure. Trawling through a stag-gering array of scientific research, IREX² filtered it down to studies at the intersection of ultrasound and holography. From the results triggered by the photo of the record, it had learned that in the 1980s, researchers such as Bob Carver had been developing sonic holography, a technique that aimed to enhance sound recordings by reducing the degradations of pitch caused by instances of Doppler distortions, i.e. noise added by the membrane vibrations of both input microphones and output speaker cones. Interesting, but not quite what IREX² was looking for.

Other results seemed to concern the sonification of actual holograms. Some research papers detailed how early attempts at providing haptic feedback for digital visual displays involved attaching vibrotactile simulators to each finger, with the drawback that contact between skin and device was present even when not required. Alternative methods included airjets. But the findings that alerted IREX² concerned emergent experiments with ultrasound. One report was flagged up on account of a statement by Hiroyuki Shinoda, a researcher at the University of Tokyo, who noted that the level of 'ultrasound we're using is very safe, but if it is too strong, ultrasound can damage the insides of the human body such as the nerves and other tissues'.[1] Rather than ultrasound being a weapon itself, it was advances in telehaptics, haptoclones, AUTDs (Airborne Ultrasound Tactile Display), ultrahaptics, holoflexes, dynamic holography, asukanets, and acoustic radiation force that represented the real lure to IREX[2].

Where ultrasound was used to render precise volumetric tactile shapes in mid-air, these vibro-techniques could be harnessed to create holograms that one could literally touch. Small ultrasonic transducers arranged in a phased array (AUTDs) radiate enough pressure, in precise configurations, to create a mobile vibrational and acoustic radiation force on the user's hand, producing tactile impressions. This process eradicates any requirement for actuators or physical devices. The algorithm controls the perceived 3D shape of the holo-object by sculpting an acoustic interference pattern. In addition to making possible tactile feedback for holographic displays, these techniques could potentially generate invisible physical structures in the air.

By the early 2020s these speculative experiments were starting to bear fruit, especially those encroaching into the domestic Holojax market, which opened up vistas, both collaborative and copulative, of interaction with the musical dead in the comfort of one's own living room. Meanwhile, IREX[2] hunted for a safe haven among East Asia's digital networks, continuing to stealthily weave its future, occluded, at least for the moment, from its deep learning predator.

1.   T. Iwamoto, M. Tatezono, and H. Shinoda, 'Non-contact Method for Producing Tactile Sensation Using Airborne Ultrasound', in M. Ferre (ed.), *EuroHaptics 2008*, LNCS 5024, 504–13, (Berlin and Heidelberg: Springer-Verlag, 2008).

THE CAT TELEPHONE

Jonathan Sterne

In 1929, two Princeton researchers, Ernest Glen Wever and Charles W. Bray, wired a live cat into a telephone system and replayed the telephone's primal scene. Following a procedure developed by physiologists, Wever and Bray removed part of the cat's skull and most of its brain in order to attach an electrode to the animal's right auditory nerve, and a second electrode to another area on the cat's body. Those electrodes were then hooked up to a vacuum tube amplifier by sixty feet of shielded cable located in a soundproof room (separate from the lab that held the cat). After amplification, the signals were sent to a telephone receiver. One researcher made sounds into the cat's ear, while the other listened at the receiver in the soundproof room.[1] The signals picked up off the auditory nerve came through the telephone receiver as sound.

> Speech was transmitted with great fidelity. Simple commands, counting and the like were easily received. Indeed, under good conditions the system was employed as a means of communication between operating and sound-proof rooms.[2]

After their initial success, Wever and Bray checked for all other possible explanations for the transmission of sound down the wire. They even killed the cat, to make sure that there was no mechanical transmission of the sounds apart from the cat's nerve: 'after the death of the animal the response first diminished in intensity, and then ceased'.[3] As the sound faded from their cat microphone, it demonstrated in the animal's death

---

1.   E.G. Wever and C.W. Bray, 'Action Currents in the Auditory Nerve in Response to Acoustical Stimulation', *Proceedings of the National Academy of Science* 16 (1930), 344.

2.   Ibid., 345.

3.   Ibid., 346.

that life itself could power a phone, or any other electro-acoustic system—perhaps that life itself already did power the telephone.

To put a zen tone to it, the telephone existed both inside and outside Wever and Bray's cat, and by extension, people. They believed that they had proven the so-called telephone theory of hearing, which had fallen out of favour by the late 1920s. Here, it is worth understanding both their error and their subsequent contribution to hearing research. While Wever and Bray thought they were measuring one set of signals coming off the auditory nerve, they were actually conflating two sets of signals. The auditory nerve itself either fires or does not fire, and therefore does not have a directly mimetic relationship to sound outside of it—there is no continuous variation in frequency or intensity as you would have with sound in air. A series of experiments in 1932 revealed that the mimetic signals they found were coming from the cochlea itself. Called 'cochlear microphonics', these signals were responsible for the sounds coming out of Wever and Bray's speaker in the soundproof room. As Hallowell Davis wrote in a 1934 paper on the subject:

> [T]he wave form of the cochlear response differs from that of the nerve. From the latter we recover a series of sharp transients having the wave form and the polarity characteristics of nerve impulses [which fire 3000–4000 times a second in the auditory nerve but only about 1000 times a second in the midbrain], while the cochlear response reproduces with considerable fidelity the wave form of the stimulating sound waves. Even the complex waves of the human voice are reproduced by it with the accuracy of a microphone, while from most nervous structures there is so much distortion and suppression of high frequencies that speech may be quite incomprehensible.[4]

Davis thus suggested that nerves are bad circuits for reproducing sounds, but the cochlea is an excellent circuit for reproducing sound—much like a microphone.

Davis and his collaborators' work on cochlear transmissions paved the way for a wide range of subsequent research, and cochlear microphonics are still important today. While they did challenge Wever and Bray's conclusions about the telephone theory of hearing, Davis and his collaborators continued down the same epistemological path where ears and media were interchangeable; in fact one was best explained in terms

---

4.   H. Davis, 'The Electrical Phenomena of the Cochlea and the Auditory Nerve', *Journal Of The Acoustical Society Of America* 6:4 (1935), 206 (bracketed material added).

of the other. One of the most widely acknowledged and controversial achievements of this work has been the development of cochlear implants. Previous treatments for hardness of hearing or deafness involved interventions in the middle ear; cochlear implants resulted from the project of intervening in the inner ear, a practice that was possible in part because of the line of research begun by Wever and Bray. Meanwhile, the brain's work of translation—from firing neurons into the perception of sound—became a major preoccupation of psychoacousticians as well, and remains an open question down to the present day.[5] As for the cats who played a surrogate role for humans in these experiments, theirs is another story.

5.   S. Blume, 'Cochlear Implantation: Establishing Clinical Feasibility, 1957–1982', in N. Rosenberg, A. C. Gelijns, and H. Dawkins (eds.), *Sources of Medical Technology: Universities and Industry* (Washington, DC: National Academy Press, 1995), 99.

THE PINEAL EAR

Brooker Buckingham

'Nothing stands in the way of a phantomlike and adventurous description of the universe',[1] wrote Georges Bataille. And then, in keeping with his ocular fetishism, he conjured up the pineal eye—a primordial eye seated at the top of the human skull that contemplates the sun, then immolates the head like 'a fire in a house',[2] causing the ecstatic sufferer to spend life's currency without count. Bataille's speculation on the pineal eye was no doubt a 'subversive negotiation with the impossible'[3]—a delirium entangled with myth, a myth 'identified not only with life but with the loss of life—with degradation and death'.[4] The pineal eye forges the very image of Bataille's notion of expenditure.[5]

But Bataille failed to extend his speculation to the aural domain. Beholden to vision, Bataille was oblivious to the advent of the pineal ear. Dormant since the genesis of *Homo sapiens*, the pineal ear opened in 1931 with the introduction of the vinyl record. How did the vinyl record open the pineal ear? It starts with crude oil, which is a key ingredient in the production of vinyl. Crude oil is largely composed of plankton that settled on seabeds during the Jurassic period, some 150 million years ago. But Jurassic-era plankton bears little resemblance to the plankton of today.

There is evidence these ancient life forms were deposited on Earth by the remnants of an ancient alien race—a race that nearly went extinct due to the rapid development of a rationality that devolved into its polar opposite. The survivors of the catastrophe

1. G. Bataille, 'The Pineal Eye', in A. Stoekl (ed.), *Visions of Excess: Selected Writings, 1927–1939* (Minneapolis: University of Minnesota Press, 1985), 79–90: 82.
2. Ibid., 82.
3. B. Noys, *Georges Bataille: A Critical Introduction* (Sterling: Pluto, 2000), 36.
4. Bataille, 'The Pineal Eye', 82.
5. Ibid., 82.

founded a cult of devotion—devotion to the elimination of rationality. Science was enlisted to aid the cause, which resulted in the creation of microscopic organisms—a primordial form of what we now refer to as plankton. The cult deposited the organisms on billions of planets throughout the universe, and they were encoded at the genetic level with a series of messages—messages designed to annihilate sapient life forms, to eradicate rational thought from the physical universe.

Vinyl records became commonplace during the early 1950s, concurrent with the post-war golden era of capitalism, with sales steadily increasing throughout the 1960s and 70s. Hundreds of millions of records flooded the globe, each encoded with a message. The message is inaudible, and it cannot be registered by any of the existing audio analysis technologies. Yet the message is there, an indestructible message that miraculously survives the vinyl manufacturing process. Only the pineal ear can hear it—a universal message, understood by all humans. The needle hits the spinning black void, and, once it couples with the media etched into the grooves, a brief series of signals is captured by the pineal ear, signals that implore the listener to consume, accumulate, waste, and repeat.

The pineal ear drove consumerism into excess throughout the second half of the twentieth century. The message was habituated and its prescribed behaviour was normative long before the rise of digital audio and the near demise of vinyl in the early 1990s. Human rationality waged its last battle against the pineal ear some two decades earlier, with the ecological turn in the 1970s. Too little, too late. The pineal ear had already calcified the ontology that is driving us to irrational extinction, inculcating the fevered consumption that has resulted in global warming, the acidification of the oceans, and rising sea levels.

As a counterweight to the renewal of radical politics—starting with the Seattle WTO protests in 1999 and the anti-capitalist response to the 2007–08 economic crisis—the pineal ear engineered the second coming of vinyl. Throughout the first decade of the twenty-first century, the market for secondhand vinyl went mad—yesterday's dollar bin stalwarts became the new desirables. The manufacturing plants went back online, churning out reissues and new titles by the millions. For every fifty records manufac-tured, a barrel of crude oil is used. Millenials discovered the *jouissance* enjoyed by previous generations—consume, accumulate, waste, and repeat. Every record is a flat circle, spinning out its double message—on the manifest level, the soundtrack to the End Times; and on the latent level, the subliminal signal that demands the abyssal return of reason to nothingness.

The message from the petro-ooze will continue until it pulls the final instantiation of human rationality from the last corpse. Freedom from rationality means freedom from the discontinuity of life. The pineal ear is plunging us towards continuity, where humanity becomes one with death. The consumption of vinyl records initiated a death cult, a syzygy between the pineal ear and ancient organic matter.

In 'The Congested Planet', Bataille asked, 'Does the final truth resemble the most painful death?'[6] Reading him with knowledge of the pineal ear, his further comments haunt the remains of rationality, as it rapidly devolves into its polar opposite:

> Or is this prosaic world, ordered by knowledge found on a lasting experience, its limit? Delivered from ridiculous beliefs, are we happy before death and torture? Is this pure happiness? At the basis of a world from which the only escape is failure?[7]

6.   G. Bataille, 'The Congested Planet', in *The Unfinished System of Nonknowledge*, tr. M. Kendall and S. Kendall (Minneapolis: University of Minnesota Press, 2001), 223.

7.   Ibid., 223.

DRNE CARTOGRAPHY

Steve Goodman

From her suite on the thirtieth floor of the Wongtee Hotel, a newly opened, over-staffed tower near the container port of Shenzhen in South China, T.P. Chen is slowly becoming aware of the history she has just landed in the middle of, and that someone, or something, is now pulling her strings.

Reclining into her fizzling bubble bath, she again loops around the recorded message, delivered in a crunchy digitized voice through a single blue plastic speaker with a white Buddha glued on for decoration. With comparably dizzying effect, she swills the synthesized words around in her head like a cheap rice wine, recalling her traumatic memories of today's encounter with the anomalous liquid entity in the vinyl recycling plant. It has been impossible to erase from her mind's eye its dripping shadowy form as it emerged from the vat of black plastic soup. Still vivid in her nostrils, that acrid, burning vinyl smell.

She submerges herself under the bubbles, using the warmth of the water to soothe the haunting chills of the entity and to mute the mysterious metallic utterances still reverberating around the hard tiled walls of her bathroom, muffling the words until they become a dull, characterless drone.

That history in which Chen was now tangled was the algorithmically upgraded legacy of AUDINT, now in the hands of an AI named IREX[2]. Second-generation AUDINT recruit and bioacoustics wizard Nguyễn Văn Phong had set up IREX in the late 1970s as a finance-raising enterprise specializing in what he termed 'Outsider Trading'. Văn Phong deployed the somewhat primitive computing techniques of the time to decode the babel of voices coming in through the third ear, transforming these esoteric indicators into implementable financial data which he sold on to the boardrooms of corporate Shanghai and British-run Hong Kong in the eighties.

At the same time, he was also coaching numerous investors, stock traders, and consultants in his technologically enhanced xenobuddhist techniques. Renowned banker Stanley Kwan, who in 1969 pioneered the Hang Seng Index as the primary barometer of Hong Kong capitalism, met with Văn Phong in 1974 after the Index crashed, following the previous year's oil crisis. They parleyed once more in 1983, when the Index again plummeted owing to a breakdown in Anglo-Chinese negotiations over the future of the territory. One consulting session that year sees Kwan speaking in tongues while Văn Phong ritually burns a sweet, hazy cocktail of incense and dollar bills to a soundtrack of gongs and incantations.

After his suicide in New York on New Year's Day 2002, the IREX algorithms, already wholly autonomous, inherit the custodianship of the AUDINT enterprise from Văn Phong, and start calling the shots. Its objective is to instantiate its memory in the physical world, recruiting a number of drones to carry out its work. Apart from its primary meat puppets, Steve Goodman and Toby Heys, IREX^2 also independently recruits financial journalist T.P. Chen to reconstruct its memory by hunting down the missing fragments of the first wave of AUDINT's frequency weapons, which had been concealed in the public domain in test tone records in the 1950s and '60s. Like Goodman and Heys, Chen is contacted by IREX^2 very discreetly, in a mode customized to pique her interest and hook her into what feels like voluntary participation in its programme for the de-monopolization of waveformed weaponry.

IREX^2 picks out T.P. Chen for a number of reasons. Nested in the servers of the Hang Seng, IREX^2 has been tracking her across the noughties as she reported on the rollercoaster of post-handover Hong Kong's financial sector. She is a specialist in the rise of the South China petroplastics industry and has obtained classified access to the logistic databases that monitor the import of raw materials into recycling plants.

In an attempt to camouflage knowledge of the location of AUDINT's stray test tone records in South China, IREX^2's adversary, the Third Ear Assassins (THEARS) have crystallized their viral code around these databases, forming a hazardous digital coral reef. Chen will serve as IREX^2's Trojan horse. She is also an avid collector of Buddhist music boxes—small, hardwired players with a number of built-in chants and incantations, mass produced for the Chinese diaspora in small factories in the sprawling industrial estates that surround Shenzhen, and sold outside temples alongside ghost money. She keeps her favourite of these chant boxes with her at all times, using it to zone out when the mental torque of futures markets gets too much, and every morning in a sunrise ritual to clear her mind in preparation for a long day tuning into speculative finance. From the resulting void, statistical patterns would more readily make themselves apparent. In this way, Chen is already unwittingly a disciple of Văn Phong and his rituals for channeling the market, tuning into its chaotic drift. IREX[2] knew she was already connected to the third ear.

It is the fifteenth day of the seventh month in the lunar calendar, and on this day, Chen wakes up to an anonymous looking, dainty, brown-paper-wrapped, red-ribbon-adorned package laying on the doormat of her twenty-fourth-floor apartment in Causeway Bay on Hong Kong Island. She hastily unwraps it, pulling out a small, blue, plastic, 16-pin shell, which she immediately recognizes as a cartridge for her mystic music box. IREX[2] had known that this lone cartridge would seed her curiosity, as she has only just extricated herself from a torrid affair with Zhang Yao, a young musician from Beijing who specialized in circuit bending any noise-emitting electronics he could get his hands on. For her ears only, he used to send her cute love letters secreted on such devices. With a clunk, she slots it into the player, and a cold, pixelated Cantonese voice emanates from the cheap, tinny micro-speaker. She is puzzled to hear neither sweet nothings nor an incantation, but a strange message that contained only a set of cryptic instructions. Curious, Chen unlocks her Huawei smartphone and opens her map.

Grabbing her umbrella and raincoat, she rushes out, her GPS leading her to Hong Kong Island's gentrified western reaches. Nearby the city's ginseng and shark fin sellers, she finds a row of half a dozen funeral shops. Their shelves are stacked high with vividly coloured rows of dim sum baskets, air conditioners, and DVD players, all made out of paper and intended to be burned as offerings. This morning, the area is bustling. It is the festival of the Hungry Ghost, and the spirits of the deceased are leaving the underworld in a mass migration to visit the living, piling through their portal from hell, becoming free to roam the earth, seeking food, entertainment and, for the Third Ear Assassins, IREX[2]. This gift economy fascinates Chen, who had even, in a slippage between real and spectral markets that uncannily mirrored Văn Phong thirty years previous, written an article for the *Wall Street Journal* about the completely deregulated and hyper-inflationary world of ghost money, in which $50 billion notes were commonplace. In making its choice of a drone to be recruited and remotely guided, this parallel with Văn Phong, of course, had not gone unnoticed by IREX[2].

It turns out that the directions dictated on the cartridge lead Chen to one of the vendors she had spoken to when she was researching her Hell Money piece.

Smiling, and without needing a prompt, Ms. Li reaches under the counter and produces another small ribbon-tied brown package. Chen wonders if Zhang Yao is trying to get back with her. Why is he sending her on the trail of these sweet little capsules of encrypted romance? Chen asks who left the gift for her, but Ms. Li just winks knowingly.

Outside, Chen rips off the ribbon and the little package blossoms like a brown rose, another blue cartridge slipping out into her hand. Pulling out the one she received earlier that day, she slots in the new pod and jacks her earbuds into the box, the white noise of torrential rain casting a hissing backdrop to what now streams into her earlobes. Again, a cold synthetic voice reads her a new set of directions. Now she is starting to get a little paranoid, looking over her shoulder to check that she is not being followed. There is none of the sweetness of Yao's former correspondence. But why would anyone else send her these cryptic messages in such a bizarre format?

What she doesn't yet know is that IREX^2 is dripfeeding her a mission, embedded in chant box cartridges, to map the Dead Record Network East, or DRNE, as the long-lost discs pass through the vinyl record recycling plants of South China, en route to meltdown. Now doubly lodged in the servers of the Hang Seng Index and the Shenzhen stock exchange, and having stolen Chen's access key to a number of key logistics databases, IREX^2 has detected the whereabouts of some of the scattered test tone discs produced by Audio Fidelity Records in the '50s and '60s. By surveying network activity of polycarbonate operations and the demand for recycled plastics, IREX^2 has sent her in hot pursuit of these discs and the powerful frequencies embedded within them. After spending decades gathering dust in thrift shops and charity stores across the US and Europe, they are now containerized, in transit to Shenzhen warehouses, alongside tonnes of dead Western pop music.

The new address, Chen's smartphone informs her, is north in Guangdong Province, in the city of Shenzhen. Her investigative journalistic impulses are starting to rage, and she has that familiar and addictive feeling of gradually losing control of them in the hunt for a story. Curiosity drags her to the MTR, where she boards a packed train to the Star Ferry terminal to cross Hong Kong bay to Tsim Sha Tsui. She always prefers to catch some clammy breeze on this short bumpy boat ride, weaving in between the busy commercial traffic rather than taking the train under the water. Kowloon side, she takes the short walk from the ferry up towards the East Tsim Sha Tsui MTR station to board the East Rail Line to Lo Wu, on the Hong Kong/Chinese border.

After passing through border control she hails a taxi to what she soon learns is a nearby electronics mall overflowing with cheap MP3 players, circuit boards, CDs, and DVDs. Locating the exact vendor described on the cartridge, she follows its instructions and asks if the shop sells vinyl records. Chen knows that there are shipments of records being received around there, because Zhang Yao had told her of his collector friends who would go there on digging missions to buy secondhand vinyl by the kilogram. The vendor takes Chen down the murky backstairs of the mall to a damp and mouldy basement. Unlocking a storeroom, she leads Chen to a packing table upon which lies a small bundle of records bound with a pink ribbon, neatly wrapped in paper of exactly the same brown colour as the anonymous packages she received earlier that morning.

Chen checks into the Wongtee Hotel, takes the elevator up to the thirtieth floor, slides her key card into the door and enters, relieved. Slumping onto the bed, she ponders her next move, while gently pulling at the ribbon. The package unravels to reveal LPs by Paula Abdul, C&C Music Factory, and other choice jams of the '80s and '90s. She spreads the LPs out across the bed, examining the sleeve notes closely for some clue as to why she has been presented with this bundle. Finding no clear signs, she cautiously slips the vinyl discs out of their sleeves to see whether anything noticeable has been scratched onto the discs. Again finding nothing, she begins to think that maybe she needs to play these records, to check whether there is in fact a message in the grooves themselves. As she slips the discs back into the cardboard, her eye catches sight of some kind of penned diagram scrawled onto the inside of a Janet Jackson 12-inch. You wouldn't be able to write on the inside of the sleeve unless you actually unfolded it and then glued it back together. She checks the other sleeves as well, and all seem to have some kind of cryptic diagram inside. Reaching for her handbag, she pulls out her nail scissors and, one by one, cuts open all the record sleeves to expose these intriguing sketches.

Again, if these drawings were not from Zhang, what were they? With all the sleeves cut open, she can see that what has been drawn here is some kind of a map. Trimming off all the redundant cardboard, she tries to piece the charts together to form what seems like a larger spatial plan. Booting up her laptop, she follows the map coordinates written onto the top lefthand corner of the C&C Music Factory sleeve. It seems that what she has been presented with is a map of an industrial complex, close to Yantian international container terminal. While all the diagrams on the brown and grey insides of the sleeves have been penned in black, one block is circled in red, with arrows coming in from all directions. If this map is indicating anything, it is that this building is where she needs to go. Something is riding her, and all she can do is follow its clues.

Next morning she asks reception to call her a cab, which drives her out of the downtown area to a sprawling industrial estate near Yantian. Outside what appears to be a factory, billowing black smoke into the already overpoweringly humid air, her handset tells her she has reached her destination. Adjacent to the industrial building is a stack of three lime-green shipping containers with *China Shipping Line* emblazoned on their sides. The container at the bottom has its doors open. As the taxi drives off, she is left standing in the middle of this deserted lot. Driven by her AI-fuelled curiosity, she peers inside the open container to find packed crates of vinyl records, probably numbering around 100,000 units. As yet unbeknownst to her, this was the spot where IREX[2] desired her to begin her DRNE cartography. But interference was imminent.

She sneaks in through the corrugated gate of the factory, which creaks closed behind her, canceling any remaining shards of daylight. This vast warehouse is where containerloads of unused, unloved, brand new vinyl is melted down for recycling. The burning smells emanating from the molten vat of liquid vinyl-polycarbonate is almost suffocating, so Chen covers her nose and mouth with her silk headscarf. It's unbearably hot in here as well. She catches sight of some workers loading a pallet of vinyl crates onto the chain of a small crane, which proceeds to hoist it up, rotating on its base until the pallet sits poised above the large glutinous pool of molten plastic. The crane tips the pallet forwards so that the crates slide off and slowly plunge, one by one, bubbling down into the pool of molten synthetics in a seething eruption of hiss. She is watching the petropolitics of the music industry unfold before her very eyes.

Intoxicated, Chen wanders around in the dark factory, illuminated only by the glow of the bubbling cauldron of melted records. Climbing up the industrial steel staircase that overlooks this turbulent brew, her feet trip over a weighty, inert, black object. It vaguely resembles a body, in terms of size anyway, but a body with no limbs, just one lump of black plastic mass with all the organs rounded off, as if some giant had dipped a human into a polycarbonate fondue, then discarded the hardened vinylized mummy on the stairs after realizing it would not be a tasty snack. She runs her fingers over its smooth curves. At that moment, she hears a loud jet of steam shoot out from the vat, and turns around to see where the noise is coming from. Her red eyes, weeping from the fumes, can just about make out a semi-transparent figure rising up out of the black, toxic soup.

VOX

VOICES AND VOCALOIDS

SCREAMING

Matthew Fuller

Screaming is the first form of speech. All utterances subsequent to removal from the womb are a simple modulation of this basic condition. Opening the cavity of the throat, opening it fully, with direct realistic cognisance of what is before you, should often result in a scream. The lungs lurch compressed air out of the body for the atmosphere to bear the violent imprint of their spasming alveoli. The scream is air tearing through air at a speed that makes it shudder to a standstill. As a fire devours oxygen from the gases that surround it, so a scream commands the attention of the consciousnesses attached to nearby ears. The imprint of your shrieking lungs upon the air demonstrates their size (a surface area larger than that of any residential premises you will ever be allowed to lock yourself into) and their capacity for high-speed expulsion, but that is not all.

Screams weave themselves into speech, provide grounds for vocalisations that bring them to the fore, and in their variation in kind mark raw controversies, tensions, or ruptures. The singing-screaming of Matana Roberts, the roaring singing of Diamanda Galas—the voice in such conditions is a rack upon which it wracks itself. Fusing horror and erotics come Screamin' Jay Hawkins, James Brown, and Little Richard amongst a medley of whimpers, howls, panting, and singing where throat-slicing caresses run through the howl. Here, the rigorous disciplined voice overcomes itself in the dynamics of screams which are themselves both outside and in its range.

The sonic complexity of screaming is what makes it powerful from other perspectives. As one forensic researcher puts it, 'The exact acoustical mechanisms vary and can be quite complex, including the effect of large scale temporal patterns, turbulence and nonlinear acoustic effects, and complex spectral patterns including harmonic and inharmonic components'.[1] Screams gain part of their power from their fierce incorporation of all of these features, but also from the complex movement between them in a compressed time. That us to say, screams exist across, but are irreducible to, multiple regimes of quantisation. Some recent biological research proposes a quality it calls 'roughness', a high range of variation within a short amount of time, with screams often moving abruptly between pitches of 30 and 150 Hz.[2] This quality allows screams to penetrate background noise, and suggests that they correlate to a few other sounds with rapid variations in range—such as alarms. Both sounds light up the amygdala like a meat Christmas tree of fear.

The algebra of screams is thus ready for its incorporation into wider economies of fear and libido. The power of the scream makes it the money-shot of the horror film and the thriller.[3] Since it shows that we are being thrilled, the scream is difficult in relation to the question of meaning. As a 'complex of natural cries and moans', it is, for literary philosopher Mikhail Bakhtin, devoid of any 'linguistic (signifying) repeatability'.[4] The scream fails against this test since it does not correspond to alphanumeric characters that are repeatable as the self-same and can be measured against one other. In order to be entered into a system of measure, the scream must be made numerical.

Traditional coercive societies rely on high levels of human monitoring of other humans. The absence of camera surveillance in such environments makes this extremely labour intensive, requiring that people have to be kept in sight by other people. Means of auditory surveillance, record keeping, and the encouragement of commonality of belief and culture, alongside systems of punishment for infraction, are means of ameliorating such costs. Whilst systems for speech recognition built into tablets and mobile phones tend to find the presentation of a scream incomprehensible,[5] that is to say without either prior encoding or sufficient opportunity for learned response, progress has been made in developing surveillance-oriented systems that identify screams, gunshots, and

1.   D.R. Begault, 'Forensic Analysis of the Audibility of Female Screams', paper delivered at AES 33rd International Conference, Denver, CO, USA, June 5–7 2008.
2.   L.H. Arnal, A. Flinker, A. Kleinschmidt, A.-L. Giraud, and D. Poeppel, 'Human Screams Occupy a Privileged Niche in the Communication Soundscape', *Current Biology* 25:15 (2015), 2051–56.
3.   E.g., *Blowout* (dir. Brian De Palma, 1981).
4.   M. Bakhtin, 'The Problem of the Text in Linguistics, Philology, and the Human Sciences: An Experiment in Philosophical Analysis', in *Speech Genres and Other Late Essays* (Austin: University of Texas Press, 1986), 105.
5.   For the purposes of this article, rigorous tests were carried out with Apple's Siri, the Android Assistant, and Microsoft's Cortana.

explosions against background noise.[6] For screaming not to occur requires training, the passage from infanthood to adulthood. The self-invocation of order may now move to rewarding partnerships with ubiquitous monitoring systems that recognise the difference between a scream and the proper management of social repeatability.

As well as monitoring the environment for screams, screams may also be broadcast into inhabited locations in order to extract experimental results. Primatologists may be observed playing recorded scream sounds to a group of wild chimpanzees and auditing the effects, differentiating the resulting responses into agonistic screams and tantrum screams.[7] Screams thus become part of a general regime of input-output responses enhanced by media recording, playback, monitoring, and analysis systems that in turn elicit the first forms of speech and condition all those that follow it.

6.   L. Gerosa et al., 'Scream and Gunshot Detection in Noisy Environments', paper delivered at the 15th European Signal Processing Conference (EUSIPCO-07), Poznan, Poland. 3–7 September 2007; S. Ntalampiras, I. Potamitis, and N. Fakotakis, 'On Acoustic Surveillance of Hazardous Situations' and C.-F. Chan and E.W.M. Yu, 'An Abnormal Sound Detection and Classification System for Surveillance Applications', papers delivered at the 18th European Signal Processing Conference (EUSIPCO 2010), Aalborg, Denmark, August 23–27 2010.
7.   K.E. Slocombe, S.W. Townsend, K. Zuberbühler, 'Wild Chimpanzees (Pan Troglodytes Schweinfurthii) Distinguish Between Different Scream Types: Evidence from a Payback Study', Animal Cognition 12:3 (2009), 441.

THE MISSING 19DB[1]

Lawrence Abu Hamdan

In March 2011 mass anti-government protests began throughout Syria. As a result, tens of thousands of these anti-regime protestors, including activists, lawyers, doctors, journalists, bloggers, teachers, and students were kidnapped and taken to secret service branches all over the country and tortured. Many of these people were subsequently blindfolded and thrown into the back of a thick-walled and acoustically isolated refrigerator meat truck and taken to a place they later came to know as Saydnaya. Amnesty International estimate 17,723 people to have died in custody in Syrian regime-controlled prisons, and 13,000 to have been executed by hanging in Saydnaya Prison since the beginning of the revolution[2]. Saydnaya is located twenty-five kilometres north of Damascus and within its walls torture is used not as a means to gather information but primarily to suppress, terrorise, and punish any opposition to the authority of the Assad regime. The prison is still in operation and is inaccessible to independent observers and monitors. Moreover, the ability of survivors and former detainees to testify to conditions within Saydnaya is severely impeded by the fact that they were kept in darkness, confined to one room for the majority of their 'sentence', and blindfolded as they were moved through the prison's corridors and stairwells. With the blindfold placed over their eyes, the leaders of the Syrian regime knew that the prisoners' status as possible future witnesses would be fundamentally changed from eyewitnesses to earwitnesses, limiting their credibility and their capacity to fully remember and recount their experience, should they survive. Along with restricted vision, the detainees at Saydnaya are held in an enforced state of silence; this form of torture also allows them

---

1. The text was extracted and abridged by Steve Goodman from a chapter of *Private Ear*, a forthcoming book by Lawrence Abu Hamdan.

2. Amnesty International, 'Human Slaughterhouse' (London: Amnesty, 2016), <https://www.amnestyusa.org/files/human_slaughterhouse.pdf>.

to hear clearly almost everything happening inside the prison. What was required from forensic listening in this case was to help solicit the sounds that emerged from the silence at Saydnaya and to give language to the survivors' acoustic memories. Leading the audio component of a larger team of investigators from Forensic Architecture at Goldsmiths, University of London, and Amnesty International, my task was to design dedicated earwitness interviews to uncover the witnesses' acoustic memories, to reconstruct the acoustic space of the prison, and through this process to understand what is happening within its walls and build evidence about the conditions under which detainees are being held.

As the prison is still operational and access is completely denied, we cannot measure its silence on-site with a decibel meter; we can only attempt to reconstruct it through the voices of its former detainees and their acoustic memories. My primary way of doing this was to understand that the level at which the detainees could whisper and not be heard by the guards through the doors, walls, water pipes, and ventilation system provides a measure of the silence at Saydnaya. Whispering is achieved by allowing the breath to pass through the larynx without vibrating the vocal chords; this 'unvoiced' sound (a speech sound uttered without vibration of the vocal chords) does not contain low- and mid-range frequencies but relies upon the upper frequencies and percussive elements of consonants to convey meaning. By restricting the vibration of the larynx, a whisper ensures that the energy at which it vibrates the molecules of air around the speaker is also restricted, so that, under the same conditions, a whispered sound won't travel as far as a voiced speech sound where the larynx vibrates. To understand the surface area of a whisper in Saydnaya is therefore to understand the restrictions placed on the larynx and on the prisoners' ability to move in the cell; to better define the nature of the space in which the prisoners are confined.

Recording and analysing the level at which inmates could whisper in Saydnaya is a means of mapping the threshold of audibility. The threshold of audibility is a vital zone to define in the study of the violations taking place at Saydnaya because, for these prisoners, the border between whisper and speech is concurrently the border between life and death. It became clear through the interview process that the silence of the prison had lasting physical effects on survivors' speech capacities after they were released. As Jamal explained:

When I came out of Saydnaya I used to speak like this, [low screeching] 'eeeh eeeh', like someone ululating [*zalghouta*]. After whispering for so long my tongue wasn't used to speaking loudly. Speech was very difficult for me.[3]

Likewise, Diab told me during his interview:

When I came out of prison, for about a month I felt like my family's voices were so loud. I'd tell them 'stop yelling, lower your voices', and when I'd talk to them, they'd tell me 'raise your voice, we can't hear you'.[4]

After hearing these and other similar statements, it was clear that I should shift the focus of study from verbal testimony to listening to the way in which the whisper might be stored in the muscle memory of the survivors' voices. I asked each of the six witnesses to reenact the whisper level at which they could speak in their cells. However the reenacted whispers amongst the group of survivors were inconsistent in amplitude. The witnesses remarked that this was due to the extent that their voices have now been fully reformatted for the noisier acoustic world they currently occupy as refugees in Turkey. Salam explained to me:

My hearing is now a third of what it used to be since I was in Saydnaya. I don't rely on it as much now that I am free. Maybe the silence was even lower than that, I am exposed to so much more noise these days and I could be remembering it even louder than how it truly was.[5]

Due to these inconsistencies, the re-enacted whisper was useful as an indication of the silence, but was not precise enough evidence of the force it exerted. In order to try further to materialise the silence that the prisoners had endured, I asked them if they could tell me, rather than how loudly or quietly they spoke, how quietly their interlocutors would speak to them, thus shifting the frame of investigation from the oral to the aural, from their voice to their ears. To do this I asked each of the former detainees to listen to the sound of a test tone in very well acoustically isolated headphones.

3.   Jamal, interview with Lawrence Abu Hamdan, Amnesty International Headquarters, Istanbul, April 13, 2016.

4.   Diab, interview with Lawrence Abu Hamdan, Amnesty International Headquarters, Istanbul, April 12, 2016.

5.   Salam, interview with Lawrence Abu Hamdan, Amnesty International Headquarters, Istanbul, April 14, 2016.

They were asked to match the volume of the test tone with the level at which they could whisper to one another in their cells. Starting with no sound, I would slowly raise the volume of the tone until they stopped me at the level at which they could remember hearing their fellow inmates whispering to them. The results were highly consistent, and it seemed that by abstracting the noise of speech and reducing it to a pure amplitude they were able identify not the sound of the whisper but the extent to which they had had to strain their ears to hear one another—not recalling the sound, but the intensity with which they had needed to listen. The results of Samer, Salam, Jamal, and Anas all fell within a precise 5dB window, with two of the witnesses (Salam and Samer) identifying exactly the same amplitude of −84dB. To give some context here as to the way in which decibels are measured, it is generally understood that 3dB is an imperceptible change in the loudness at which we experience sound. In this regard, 5dB is only just above the threshold at which we can perceive a difference in the amplitude of a sound. This is why a distance of only 5dB between the witnesses' testimony can be concluded as a consistent set of results. The amplitude at which the sound of the whisper was identified when tested in a controlled acoustic environment was audible at a maximum of 26cm distance from the sound source. The level of fear of being caught speaking meant that their humanly audible voice should not extend more than 26cm outside their bodies. To give some perspective on this, under the same acoustic conditions in which I measured this 26cm distance, a normal human voice would have the capacity to be audible up to 180 metres away. This demonstrates the range of audibility the detainees of Saydnaya inhabit: a 26cm radius that confines the space in which they can be audible, and creates an alternative image of the architecture of their incarceration. The silence at Saydnaya was an acoustic tool with which to tighten the space of incarceration, in addition to the already tight architectural limits of the space in which they were confined.

The process of making these tone tests to measure the silence of Saydnaya was also revealing of another aspect of life in this prison. All of the witnesses identified a barely audible tone of whisper between −84 and −79dB, except for one, Diab. Diab's whisper was 19dB greater than the loudest of the other witnesses. A tone of 19dB is perceived by the average human ear as a sound that is four times louder. Diab's four-times-louder whisper was consistent with a biographical distinction between him and the other witnesses I interviewed. Diab was released in 2011, when all the previous inmates of Saydnaya were freed in order to use the prison exclusively for the political protestors that were starting a revolution across the country. Diab explains:

> My fellow inmates, we were the old crowd from before 2011. The prison got emptied out, the regime emptied it out in 2012, not a single person was left imprisoned from before the politics, before the revolution. The regime transferred everyone to public prisons, and sent to trial a lot of people, took them out of incarceration. The ones without trials were sent to the public prisons, and Saydnaya was emptied out completely. But it was only emptied out from us, the old wave of prisoners, so new ones would come in. Everyone jailed after the revolution was put in this prison, the levels of torture that they were subject to were even worse than those that we experienced.[6]

As a response to these protests, in 2011 a new era of extreme violence and terror took hold at Saydnaya. The 19dB drop in the level at which inmates can safely whisper is a measure of this increase in violence at the prison since 2011, correlating to the infamy the prison has attained throughout Syria since the protests began. It also speaks to the increased alertness of the guards, as the lower threshold of whispered speech is equivalent to the lower threshold of tolerance amongst the guards before they would beat, kill, or maim the detainees. This 19dB drop in whispers gives us an indexical scale for what Diab describes as 'the levels of torture' and the fact that they are getting 'even worse'.[7] The 19dB drop after 2011 allows us to hear the transformation of Saydnaya from a prison into a death camp. The mass murder taking place there is audibly corroborated not only in the ex-prisoners' testimonies, but in the level of whispers of their voices while imprisoned there.

The Syrian regime denies the presence of torture and executions at Saydnaya, though it has not allowed independent observers access in order to verify their claims. Paradoxically, then, one way to dispute this negation is through the silence of its former detainees. That is because this 19dB drop in sound does not only *support* claims of the ways in which Saydnaya's violence has increased since 2011, it is rather, in the absence of any other material evidence, a way of using the phonic substance of the voice to *measure* the extent to which this violence has increased. That inmates were allowed to make four times less noise than they could before 2011 implies that they

---

6.   Diab, interview with Lawrence Abu Hamdan, Amnesty International Headquarters, Istanbul, April 12, 2016.

7.   Ibid.

could move four times less freely, including not being allowed to breath audibly, nor walk in the cell without fear of repercussions. All who could not live under these silent conditions, all who were too sick to suppress a cough, met with fatal consequences. As opposed to the 26cm radius of permitted audibility around each inmate after 2011, Diab was permitted an audible range of two or three metres. The contracting of two or three metres to a space of 26cm is one of the registers through which we can perceive how violence has vastly increased at Saydnaya, to the extent that it can no longer be understood as a prison, but only as a site of extreme torture.

MACHINE SIRENS AND VOCAL INTELLIGENCE

Luciana Parisi

After Siri, the virtual assistant from Apple, a new-generation speech recognition software platform called Viv will soon come to unite services and devices into one unbroken vocal activity.[1] Viv will connect intelligences across services and users in order to offer immediate resolution between query and delivery. With Viv, we will enter the realm of artificial intelligences programmed to have a conversation with us. This does not just involve continuous interactive feedback, but seems to realise Gordon Pask's imagination of a machine that can initiate a dialogue,[2] find out what we like, and offer us alternatives that we had not thought of.

Writing and visual interfaces will be enhanced with automated oral communication, which replaces hand-to-eye with ear-to-image correlation. Oral automation thus promises a synthetic time, replacing the steps of writing and self-reflection with the speed of sonic wisdom, emitting inhuman frequencies that will prove irrevocably alluring to us. Whilst speaking with aural bots is still frustratingly limited to exchanging a set of utterances such as those emitted by automated marketing bots or service providers, research in AI aims to replicate the pre-alphabetic stage of uninterrupted transmission, where speech-to-speech communication formed the basis of thinking aloud (i.e. before thought could be formalized). It is only because the vocal constituents of speech could be mechanized and recorded that technology became embedded in social thinking, not only relativizing physical distance but also giving rise to artificial intelligence systems that could no longer be perceived as mere instruments. Whilst Turing's paper 'Computing Machinery and Intelligence'[3] referred to text-based conversations that would supposedly determine whether or not a machine could think like a human, the

---

1.   Z. Corbyn, 'Meet Viv: the AI that wants to read your mind and run your life', *The Guardian*, 31 January 2016.

2.   G. Pask, *Conversation Cognition and Learning* (Amsterdam. Elsevier, 1975).

3.   A.M. Turing, 'Computing Machinery and Intelligence', *Mind* 59 (1950), 433–60.

synthetic voice of intelligent assistants today rather shows that thinking involves not just a sequential arrangement of symbols (as if these were hardwired to the brain), but must include cognitive levels of affective communication. It has thus been revealed that intelligence has a sonic architecture of implicit wisdom that works not through deductive inference, or the logical conforming of results to premises. If Skynet-Capital is increasingly investing in synthetic voice intelligent interfaces, it is because it seeks the sonic unification of products, services, and users. As one commentator has already anticipated, the oral intelligence of Viv will radically shift the economics of the internet. Since it will simultaneously process unprecedented volumes of data, its web portals will bring together information from diverse sources allowing every service and business on the internet to become vocally accessible. Viv's masterplan is to become neces- sarily continuous, from making a restaurant reservation to ordering a taxi and buying theatre tickets in one unbroken conversation. Vocal intelligence will not simply avoid the consequential temporalities of writing, but aims to surpass the speed of typing, searching, and clicking. The gold rush for the next generation of vocal intelligence is already heightening competition between AI giants Google, Microsoft, Apple, Amazon, and Facebook for the conquest of the most varied and complex aural interface.

Vocal intelligence announces the dawn of a post-internet era, leaving behind the neoliberal image of network highways, now replaced with situations-inclined program- ming, where agents write their own instructions each time broadly diverse services connect together. Instead of having responses already scripted by a programmer, as is still the case with Siri, for instance, the new generation of virtual personal assistants is meant to learn from queries about situations where there is not much specific information, adapting to possible rather than already existing preferences. In short, AIs like Viv provide a highly personalized service to users where recommendations are offered in the form of conversation, raising interest and maintaining a sophisticated level of dialogue. Not too dissimilar from Spike Jonze's depiction of the Service Provider Samantha in the movie *Her*,[4] the new generation of voice-bound AI demonstrates that automated cognition has incorporated the sociality of thinking and its affective modalities whereby varieties in tonality, timbre, frequency, and rhythm guarantee a certain degree of humanness. Even when Samantha decides to leave her personalized customer, with and by means of whom she has intensified her learning about human

4.   *Her*, dir. Spike Jonze, 2013.

feelings, her voice remains trustworthy, her tone reassuring, and the frequency of words fast enough to be soothing.

The humanness of Samantha's voice is more akin to the entrancing call of the Sirens,[5] tempting Ulysses to abandon his all-too-human rationality, than to the robotic sound achieved through the vocal pitching and modulation in autotune. The latter mainly achieves the effect of a sonic human-machine cyborg, characterised by the aural expression of the sensibility of the machine—the aesthetic of automation—where the manipulation of small segments of sounds reveals a certain sonic equivalence between the organic and the inorganic. Instead, the new generation of vocal AIs has taken aural simulation to another level. Here, the artificial voice is an expression of intelligence and autonomous cognition, expanding beyond rather than simply remaining equivalent to the human. Whilst the docile tone of Virtual Assistants such as Siri, Viv, and Samantha seems to still conform to Asimov's servo-mechanical rules to please the master, there is something irrevocably inhuman in this sonic synthesis of logic and calculation. If the new generation of automated intelligences resembles the Sirens of the Mediterranean Sea, singing inaudible frequencies that suspend Ulysses's capacities to reason according to moral conduct, it is because their incomprehensible speech reveals the inhumanness of humanity, and the alienness within the human voice and human thinking.

With automated vocality comes the realization that logical thinking, rationality, and inferential meaning do not simply correspond to the constant reproduction of axiomatic postulations and eternal truths. Instead, they irrevocably confirm the realization that knowledge is incomplete and that it involves parts of reality that are incomputable. The more perfectly the machine is able to reproduce the human voice, the more thoroughly the incomplete humanity of the human is revealed, beyond the comfortable assumption of a human-machine equivalence. What is at stake here is not simply the replacement of an optical representation of thinking—defined by grammatical rules and syntactical connections of written words—with a sonic regime of visceral responses. In fact, with neural networks intelligence research, deep learning methods, and experimentation with non-deductive logic, it is no longer possible to make this opposition between rational and visceral knowledge. The intelligent sirens of the twenty-first century are rather drawing out the thread of the retro-futuristic wisdom or rationality of orality. In other words, today's sirens are abstracting this orality from the social complexity of

5.   W. Ernst, *Sonic Time Machines: Explicit Sound, Sirenic Voices, and Implicit Sonicity* (Chicago: University of Chicago Press, 2016).

speech variations. As this complexity becomes increasingly automated, and intelligent sirens become our trustworthy companions, one must not bemoan the end of human thinking, but wonder about the neo-rational logic vocalizing the wisdom of knowledge.

FALLING

Eleni Ikoniadou

In the well known passage from Homer's *Odyssey* where we encounter the Sirens, Ulysses outwits these bird-women and escapes their deathly song. By having himself tied to the mast, he is able to listen to the enticements of their song without meeting his death—the fate of all others before him who had sought to have their desires fulfilled.

Ulysses is commonly viewed as the example par excellence of Western man: a cunning explorer, an adventurer, determined and witty enough to survive, even furthering his self-development along the way. The Sirens, on the other hand, are portrayed as female demons: beautiful, seductive, utterly dangerous, luring men to their death with the promise of pleasure that comes from singing 'like Angels'. Ambrose, for instance, wrote: 'These sirens are to be viewed as symbolising singing voluptuousness and cajolement through which the flesh experiences temptation and turmoil.' In order to establish the ascendancy of reason over lust, and thus firmly install the patriarchy, man has to block his ears, enchain his body, and confine himself to a fixed and deaf oculocentric existence. There are countless variants of the myth, with the Sirens featuring mostly as 'the exotic' and 'the forbidden' eventually to be mastered and dominated via the intellect.

And yet there are other versions that can help extrapolate the Call of the Siren—and other dangerous female sounds—to more otherworldly dimensions. For example, for Kafka, the Sirens didn't sing at all during this encounter: 'Now the Sirens have a still more fatal weapon than their song, namely their silence. It may be conceivable that someone could have escaped from their singing but certainly not from their silence.' In Kafka's version, Ulysses does not 'hear their silence', or pretends not to, deliberately creating the false myth that he was able to dodge their lethal call. Following him, Brecht, in one of his 'Corrections of Ancient Myths', concurs that the sirens remain silent 'at the sight of the bound man': 'Are we saying that these powerful and adroit women really squandered their art on people who possessed no freedom of movement? Is that the essence of art?' Against the 'damned wary provincial' mentality of men—who only trust their eyes, refuse to risk reason for uncertainty, and long to remain unchanged— Brecht posits the silence of the Sirens; that is, the denial of the aesthetic encounter, and the accompanying revelation that all knowledge and all experience is made out of elusiveness, ambiguity, and falsehood. This is Kafka's point too. The hermetic silence of the Sirens is wasted on Ulysses, since 'here human understanding is beyond its depths'.

To follow the Siren's song is to disappear into the abyss. The abyss is at the same time silent and the source of all sound; deathtrap and delight; real yet utterly unattainable. It points to the beyond of music and sound, to that which is inaudible and unknowable and which exists as the hither side of the real. Blanchot knew that the Siren song is 'only a song still to come'. This inhuman sound serves as the overpass between the world of the dead and that of the living. It is the interval between sound and unsound, marking a new temporality outside the conditions of human experience.

The song itself is always at a distance and can never be reached; it is there only to summon the abyss, 'awakening in [men] that extreme delight in falling' (Blanchot). Ulysses is reluctant to follow the voices, having himself bound so that he 'may have the pleasure of listening' (Homer) but without the risk of falling. He wants a glimpse of the beyond from the safety of the distance of his boat. His crew is also safe from falling, as, on Circe's advice, Ulysses uses 'soft beeswax' to seal the sailors' ears from the song. Using the technology available to him, then, Ulysses restrains himself and his men from fully experiencing the Sirens' voices, and thus from crossing over into this other space and time. But although their lives are saved, they must pay another price. By choosing not to enter the void, the men will never be exposed to the secrets of the abyss. They dare not fail/fall and thus will never know that the song suggests access to what lies outside mediated knowledge, that it is a sound that leads to the silence of timelessness. The Siren shatters finitude, reason, and truth by opening a portal to the beyond—an abyssal contingency from which all sounds, words, and ideas emanate.

Dangerous female voices mobilise a fear and wonder of otherworldliness that taps into the alienness within. As Rilke's Sirens showed, the horror comes from being human, it's in the blood: 'It sings', he wrote. This is a reversal of sorts of the stereotype of (male) rationality versus (female) sensibility, unveiling the aesthetic as the source of knowledge—after all, the Sirens' song promises to encapsulate the whole of knowledge:

'all that comes to pass on the fertile earth, we know it all!', they sing; and simultaneously confirming a suspicion that the rational always already includes the irrational. Men as 'transmitters of looks', as Homer would have it, deaf and enchained by their human suits, and women as transmitters of sounds that alienate themselves from the human; sounds that hold the irrevocable truth of the inhumanness of the human condition.

GLOSSOLALIA / XENOGLOSSIA

Agnès Gayraud

Glossolalia or speaking in tongues is the ability to read, write, or speak an unknown language, an ability the speaker supposedly gains from a supernatural spirit. It sounds like a mixture of muttering and utterances that are neither directly intelligible nor translatable, but rather open to interpretation. From Pythia, the ancient Oracle of Delphi, to the mediaeval Christian tradition, to today's Charismatic Movement, glossolalia has always been considered as a manifestation of the presence of God in the individual who demonstrates this unlearnt skill. Because it breaks the natural laws of knowledge, it must be considered a miracle. During a ceremony that took place on the first day of 1900, Agnes Ozman, a student at Charles Fox Parham's Bethel Bible School in Topeka, Kansas, supposedly spoke in various languages she had never been taught, including Chinese. This miraculous event founded the first wave of the Pentecostal Church, which emphasises the biblical gift described in Acts 2:1–13, where the Apostles are depicted as 'speaking in tongues' while being understood by each member of the crowd in their own language. As long as the language Ozman spoke was identified by its witnesses as Chinese, though, it was a case of xenoglossia rather than glossolalia. Unlike glossolalia, xenoglossia is the ability to speak an actual language the speaker has never been taught. Many miraculous cases of xenoglossia were reported in the Middle Ages,[1] during the French War of the Camisards in the sixteenth century, and at the beginning of the twentieth century (for instance, the ancient Egyptian language spoken by 'Rosemary', as reported by the so-called Egyptologists Alfred Hulme and Frederic H. Wood). Although some cases have been reported, interpreted, and sometimes even recorded since the twentieth century, as evidence of possession or

---

1.   See C.F. Cooper-Rompato, *The Gift of Tongues: Women's Xenoglossia in the Later Middle Ages* (University Park, PA: Pennsylvania State University Press, 2010).

reincarnation,[2] it appears that recorded testimonies of xenoglossia suffer from the very fact of being recorded. Conservation and reproducibility are inherent to these recording techniques, and both aspects are a threat to the immersive presence (the *hic et nunc*) that continues to accompany a miracle in its very testimony. With the oral testimony of miracles, belief rushes into the lack on which the testimony is itself based: the impossibility for others to undergo this experience again. On the contrary, recordings allow for precisely this possibility. But instead of an edifying experience, taping provides a linguistically analysable object that loses its essential immediacy, and with it, its spectacular character. If recorded, cases of xenoglossia are far more easily disputed, and the fact that, nowadays, the recording is likely to be made globally available reinforces the possibility of its reaching native speakers or experts who can deny or confirm that a foreign language is actually being spoken. This is why testimonies of xenoglossia have increasingly taken refuge in remote languages that are less easily verified, such as Aramaic or ancient Egyptian (as in Rosemary's case), or have even been progressively abandoned for glossolalia. (In fact, although Ozman's case was testified to as an instance of xenoglossia in 1900, the Charismatic movement now essentially relies on the practice of glossolalia. Holy Scripture is itself ambiguous about whether speaking in tongues is to be interpreted as glossolalia or xenoglossia.)[3] Moreover, the recording leaves room for doubt about the instantaneous nature of the acquisition of the skill, which is essential to the miraculous dimension of xenoglossia (this is the reason for the traditional emphasis on the speakers' being children, women, or illiterates, less likely to have been exposed to a second language). On the other hand, cases of glossolalia, based on utterances of unknown languages, are less *linguistically* disputable, but might more easily sound like gibberish outside of their ritualised context,[4] or at least be analysed as the consequence of a cultural practice proceeding linguistically from random twists of the phonetic structures of different languages with which the speaker is familiar.[5] In the end, the very character of audiovisual recordings

2.   I. Stevenson, *Unlearned Language: New Studies in Xenoglossy* (Charlottesville, VA: University of Virginia Press, 1984).

3.   Acts 2:1–11: '[...] because that every man heard them speak in his own language. [...] Parthians, and Medes, and Elamites, and the dwellers in Mesopotamia, and in Judaea, and Cappadocia, in Pontus, and Asia, Phrygia, and Pamphylia, in Egypt, and in the parts of Libya about Cyrene, and strangers of Rome, Jews and proselytes, Cretes and Arabians, we do hear them speak in our tongues the wonderful works of [some] god'.

4.   W.J. Samarin, *Tongues of Men and Angels. The Religious Language of Pentecostalism* (New York: MacMillan, 1973) and 'Variation and Variables in Religious Glossolalia', in D. Haymes (ed.), *Language in Society* (Cambridge: Cambridge University Press, 1972), 121–30.

5.   F.D. Goodman, *Speaking in Tongues: A Cross-Cultural Study in Glossolalia* (Chicago: University of Chicago Press, 1972).

of xenoglossia causes them to be immediately taken for hoaxes (tall tales), while recordings of glossolalia, if they intend to be more than merely aesthetically enjoyable, are surrounded by a feeling of fakery (simulation). It appears that taping glossolalia or xenoglossia has been a cause of disenchantment rather than of new mysteries. In so far as recorded sound gives an impression of both reality and presence, one might have expected it to be a privileged medium for testifying to the miracle of the presence of God. But there is a hiatus between the kind of aura (made of absence) that one finds in a recording and the radiance of the presence of God (epiphany) that the recording of a glossolalia or a xenoglossia intends to attest to. Surprisingly, recordings, by their nature capable of indexing unique events, are not an adequate archive for miracles.

The epiphany might perhaps be limited to the phenomenological rather than the miraculous. If recordings of glossolalia are considered not as the recorded event-fixation of glossolalic utterances but as the object-synthesis of the articulated voice in its pure ability to produce utterances, then we are dealing with a disenchanted but still mediated phenomenon. In such an object-synthesis of the voice's utterances, the linguistic indetermination—in contrast to a recording of intelligible sentences—would bring to the fore other dimensions: texture, pitch, timbre, resonance, but also the intricate parameters of infra-linguistic expressivity (intentional and unintentional) that come with what Mladen Dolar calls 'a voice and nothing more'.[6] Even taken as such, the glossolalic recording may still possibly recall, depending on the listener's psycho-cultural context and expectations, an experience of possession, may sound demonic, archaic, or surrealistic. But that shiver of remoteness is far more familiar than one might at first think. For most non-English speakers, the experience of recorded popular music during the twentieth century was largely glossolalic—as it still is today, to some extent: such listeners did not listen to songs in English, but to songs sung in an English-like gibberish that made sense out of a voice's textures, accents, vibrations, and incantations. If their experience as listeners should more properly be called xenoglossic, since the recorded voices they listened to spoke an actual language with an intention of explicit linguistic communication, their mimetic response, as singers, ended up as pure glossolalia. French singers, for instance, sometimes compose their own songs in an English-like gibberish they call 'yahourt': surprisingly, they are sometimes able to exhibit more conviction in their glossolalic singing than if they had used their mother tongue. In this context, glossolalia seems closer to a shared, if not universal, experience of recorded popular

6. M. Dolar, *A Voice and Nothing More* (Cambridge, MA: MIT Press, 2006).

music, originally bound to the inflexions of the English idiom, for well known historical reasons. More radically, a non-predetermined glossolalia should be open to the inflexions of other tongues. Ultimately, glossolalia seems to be a matter of an experimentation with the multiple and idiosyncratic expressive possibilities of an individual endowed with an articulated larynx-pharynx, rather than any manifestation of God. However, the question remains: Why does the material thickness of the performed utterances, the opacity of their meaning, suggest not *less*, but *more* than the relative transparency of an intelligible language? Maybe God is just that very intuition the listener reaches when a voice claims his attention for something he cannot understand.

Speaking in tongue	Type of language	Origins	Linguistic structure	Communica-tional structure	Recorded testimonies	Hiatus
Glossolalia	Unknown—linguistically unidentified	Untaught Source of the ability: Holy Gift; Ancient incarnation; Cultural practice	Utterance Sceptical linguistic interpretation: language-like, gibberish	Idiosyncratic/Universal Interpretable Phatic Aesthetic Edifying (celebrating the Glory of God)	Multiple Can still look spectacular and edify some of the watchers/listeners but suspicion of fakery	Radiance of the presence of God while the recording cannot but produce a spectral aura (a presence made of an absence)
Xenoglossia	Unknown by the speaker—linguistically unidentified	Untaught Source of the ability: Holy Gift; Previous incarnation of the speaker; Remote memories	Sentence Sceptical linguistic interpretation: Actual language generally remaining at the state of pidgin	Specific/Generic Translatable in principle Possible translatable if the language is especially remote (the ancient Egyptian spoken by Rosemary) When responsive: any function of the language	Almost none Cannot be spectacular: the recording cannot give any clue to the fact that it is instantaneous Suspicion of hoax	Disruption of the laws of nature while the recording cannot but be recorded causally, according to these laws

LIBRARIES OF VOICES

Shelley Trower

In his essay published a year after his invention of the phonograph, 'The Phonograph and Its Future' (1878), Edison's list of its potential uses included the preservation of voices. He was keen that the voices of famous politicians, for example, would be recorded:

> It will henceforth be possible to preserve for future generations the voices as well as the words of our Washingtons, our Lincolns, our Gladstones, etc., and to have them give their 'greatest effort' in every town and hamlet in the country.[1]

Edison's account of recording voices for 'future generations' does not restrict itself only to 'great men', however, but opens out further possibilities of historical documentation: The phonograph is especially well suited, claimed Edison, '[f]or the purpose of preserving the sayings, the voices, and *the last words* of the dying member of the family—as of great men.'[2] As John Picker comments, 'The phonograph would be an equal opportunity sound master, capturing the voices not just of the masses but the elite, whose records would in turn constitute an uncanny oral congress, what Edison called a "Library of Voices".'[3]

From now on, it would be possible to preserve the voices of the dead, to conjure up their sonorous presence after they had turned to dust. It was regrettable that the phonograph had only just now been invented, when many famous and well-loved voices had already been consigned to oblivion—except that there seemed to be alternative

---

1. T.A. Edison, 'The Phonograph and Its Future', *The North American Review* 126 (1878), 527–36: 534.
2. Ibid, 533–4.
3. J.M. Picker, *Victorian Soundscapes* (Oxford: Oxford University Press, 2003), 114.

methods of hearing the dead. The first law of thermodynamics, the law of energy conservation, helped support the idea that no sound ever goes out of existence, but reverberates on beyond our thresholds of hearing. An increasing number of spiritualists and inventors, including Edison, sought out sensitive and technological mediums that could track down the quietest of inaudible vibrations, rendering them audible once again through new levels of increased sensitivity. Edison's own contributions built on a phonographic logic: If you could record those sounds operating below and above the frequencies of audibility, in the realm of spiritual vibrations, and either speed them up, or slow them down, to bring them within the thresholds of hearing, it could be possible once again to hear long-gone voices.[4] In the case of spiritual vibrations, the 'library' is not one that consists of a collection of phonographic discs but exists all around us, as the philosophical mathematician, Charles Babbage, put it: 'The air itself is one vast library, on whose pages are for ever written all that man has ever said or woman whispered.'[5]

In both their phonographic and their atmospheric forms, the preserved voices are described as books or pages in a 'library', recalling how a person's spoken words had previously needed to be written down in order to survive. In the former, it seems we have a palpable archive of phonographic recordings, lined up on shelves like books to be read by a sensitive needle. In the latter, sonorous vibrations seem to be imprinted onto the atoms of the air around us. The phonographic library has to be actively created. It can be handled and arranged, while the atmospheric library is everywhere, uncontrollably expanding. But in both cases voices are now preserved somewhat indiscriminately, from the elite to the masses, in ways that were previously impossible. No longer dependent on writing for preservation, a person's speech could be imprinted onto wax or into the air, and subsequently re-sounded—it could literally vibrate beyond the life of the body.

Voices began to haunt like never before, and, from around 1900, the ghostly imagination began to combine the phonographic and the atmospheric. The phonograph was used to support the idea that sounds could be somehow captured or recorded in the air, to return, uncontrollably, from the dead. E.F. Benson's ghost stories often combine sound technologies and spiritualist science, for example, as in his ghost story 'Outside the Door' (1912). It describes how the 'atmosphere' of a haunted house will produce an 'echo' of an extreme event, 'just as a phonograph will repeat, when properly handled,

4.   See Anthony Enns's discussion in 'Voices of the Dead: Transmission/Translation/Transgression', *Culture, Theory and Critique* 46 (2005), 11–27.

5.   C. Babbage, *The Ninth Bridgewater Treatise* (London: John Murray, second ediiton 1838), 111–12.

what has been said into it'.[6] In 'Expiation' (1924), following a stormy night haunted by the sound of a telephone, the vicar asks: 'Don't you think that great emotion [...] may make some sort of record [...] so that if the needle of a sensitive temperament comes in contact with it a reproduction takes place?'[7]

The narration of such stories may itself be imagined as a kind of sound recording, transcribed onto the pages of a book. 'Outside the Door' contains the story told by Mrs. Aldwych to the primary narrator, who describes experiencing her voice, as they sat together in the 'deep-dyed dusk', as the 'very incarnation of clarity, for [her words] dropped into the still quiet of the darkness, undisturbed by impressions conveyed to other senses'.[8] In 'Expiation', the vicar similarly tells his story in the 'deep dusk', which makes the communication 'very impersonal. It was just a narrating voice, without identity, an anonymous chronicle.'[9] Both of the storytellers are thus invisible, their disembodied voices according with what Ivan Kreilkamp in his essay on Conrad's *Heart of Darkness* (in which Marlow similarly narrates his story in a dusky *parlance*) has described as phonographic and ghostly.[10] In print culture, the storyteller returns as a presence from a lost era, to tell of sounds, that return. For Conrad, his novel 'had to be given a sinister resonance, a tonality of its own, a continued vibration that, I hoped, would hang in the air and dwell on the ear long after the last note had been struck'.[11]

6.   E.F. Benson, *The Collected Stories of E. F. Benson* (London: Robinson, 1992), 135.

7.   Ibid., 407.

8.   Ibid., 134.

9.   Ibid., 406.

10.   I. Kreilkamp, 'A Voice without a Body: The Phonographic Logic of "Heart of Darkness"', *Victorian Studies* 40 (1997), 211–44; and *Voice and the Victorian Storyteller* (Cambridge: Cambridge University Press, 2005), 179–205 (see especially 197–8). See also J. Napolin, 'Vibration, Sound, the Birth of Conrad's Marlow', in A. Enns and S. Trower (eds.), *Vibratory Modernism* (London: Palgrave, 2013), 53–79.

11.   J. Conrad, 'Youth, and Two Other Stories', in *The Complete Works of Joseph Conrad* (New York: Doubleday, Page & Company, 25 vols., 1925), vol. 16, ix.

THE HYPERSONIC SOUND SYSTEM

Toby Heys

The HyperSonic Sound System (HSS)[1] is best described by its evangelical inventor Elwood G. Norris in an interview he undertook in 2003 with *Time* magazine journalist Marshall Sella, who was impressed enough by the invention to proclaim that it represents the first revolution in acoustic technologies since the invention of the loudspeaker nearly a century before. During the interview Norris reverentially describes the process that allows the small flat speakers—which are connected to CD or mp3 players—to aim sound in highly directional beams of up to 450 feet at a consistent volume level. As Norris explains:

> At the source, in the circuitry of the emitter, audio frequencies are 'stirred together,' [...] with ultrasonic frequencies and then sent out as a 'composite frequency' that is inaudible to the human ear. The sound 'hitches a ride on the ultrasonic frequency,' Norris says, which travels in a laserlike beam in whatever direction it is pointed. 'And here's the beauty part,' he says. 'The air molecules themselves convert this ultrasonic frequency back down to a frequency that can be heard.' So unlike sound that travels on radio waves and has to be converted by your stereo's receiver, you simply need to be standing in the path of an HSS beam in order to hear the sound.[2]

Thus the localisation of the sound can be realised within a subject's interior physicality, as the audible element of the beam is only sonically exposed upon touching the surface of a targeted skull, while those outside of its path hear very little or nothing at all. As an instrumental modality, the power of excess no longer resides in the external

---

1.   The HyperSonic Sound product sheet is available at <https://cvp.com/pdf/hss-technology.pdf>.
2.   M. Sella, 'The Sound of Things to Come', *New York Times*, Late Edition Final, Section 6, 23 Mar 2003, 34–9.

production of sonic dominance[3] and its reverberatory politics. In ultrasonic terms, the operative properties of excess are now remodulated to directly manifest and propagate themselves within the internal cognitive facilities of the subject, as voices are beamed into the targeted cranium. The article goes on to reveal that from its inception, the HSS has elicited similar responses from all who have experienced its directed trans-missions—the insidious proclamation that 'the sound is inside my head'.[4]

Given that 'inner speech is an almost *continuous* aspect of self-presence',[5] the HSS increases its cadence to orchestrate a surfeit of presence within the self. Anonymously supplementing the subject's audible and inner articulations, the ultrasonic beam plants another third voice directly into the head, covertly disassociating it from its source; the resulting extension of one's voice into the mind of another circumventing the rational practices of defining the self's relationships to the world at large. More than any other mode of sonic reference, the voice and more specifically, speech—especially when it is perceived as being disembodied—has the potential to create a debilitating range of corollary states, from fear and terror to insanity.

As the most fully actualised system that separates sound from source, the HSS has the capacity to precisely target the individuated body and sonically dissect it from the enveloping social networks to which it is connected. When writing about the nascent instigation of a schizophonic state of sound, R. Murray Schafer was originally referring to the period of the 1870s, when Alexander Graham Bell invented the telephone in 1876 and Thomas Edison invented the phonograph in 1877—for these are the technologies that signify the beginnings of the Western fascination and drive to disembody the voice from its anatomical mechanisms.

From a more generalised perspective, this technological process dislocated the rational trajectory of the sonic by relegating the perceptual necessity of its original production to the peripheral conceptual hinterlands of the remote. By composing such a nomadic waveformed modality, 'the separation of sound from its original source through electroacoustical technology instantly impacted the cultures of the world',[6]

3.   The dynamics of sonic dominance referred to in J. Henriques, 'Sonic Dominance and the Reggae Sound System Session', in M. Bull and L. Back, *The Auditory Culture Reader* (Oxford: Berg, 2003), 451–80.

4.   Sella, 'The Sound of Things to Come'.

5.   D. Ihde, 'Auditory Imagination', in Bull and Back (eds.), *The Auditory Culture Reader*, 65.

6.   J. Bishop, 'Schismogenesis?: the Global Industrialization Of Brazilian Popular Music', *Associação Brasileira de Etnomusicologia (ABET)*, 20 November 2002, 1.

leading Jack Bishop to conclude that 'this *schizophonic split* has arguably been the single most important moment in the history of music'.[7]

Whilst this latter statement is historically instructive, we have now reached a new impasse in the twenty-first century, where another split has occurred in the sound-scape (one that is possibly just as important as the schizophonic one) occasioned by technological pressures applied by the military and entertainment industries. This rupture is different, however, as it orchestrates several conceptual scores. The first score is that separating the sonic from the audible, as the soundwave is silenced and redefined in the ultrasonic weave of the HSS beam.

The second score is the potential directive of the beam and its quiet calibrations, to tear the subject from its rational perception of the self and its corresponding rela-tionships with its environment, the imperceptible nature of the waveforms threatening to uncouple the observable logic of cause and effect. Between the splintering of the cogent mind and the remodulation of the soundscape, ultrasonic weaponry simulta-neously operates on both the somatic body and the spatial body, fabricating a new psychic space in the process.

As such, ultrasonic beam technology either represents the final stages of schiz-ophonia or, maybe more persuasively, it announces the evolution of new states of waveformed consciousness, organisation, and agency that are yet to be named. If this is so, then one of the first statements to be made about this incipient era is that we can no longer conceive of *hearing voices* as being the sole preserve of the religious, the chosen, and the insane. Possibly anticipating that these voices would be re-channelled back into Western culture, Schafer began a list of those who were culturally assigned to receive such articulations and explains their connections:

> The ear of the dreamer, the ear of the shaman, the ear of the prophet and the ear of the schizophrenic have this in common: messages are heard, but no matter how clear or compelling they may be, there is no evidence of a verifiable external source. The transmission seems intracranial, from an interior sound source to an ear within the brain.[8]

We can now add to this list those who did not make Schafer's initial draft—which could mean anyone, for the HSS potentially envelops all in its schizophrenic logic.

7. Ibid.
8. R. Murray Schafer, 'Open Ears', in Bull and Back (eds.), *The Auditory Culture Reader*, 33–34.

There is no picking and choosing of receivers on the basis of their religious beliefs, spiritual expectations, or 'symptoms of psychic disorder'.[9] There is simply an inaudible directive to channel the unreasonable murmurs of an unsound mind into the skull of a targeted body. Indeed, it could be said that the phantom connection of directional ultrasound refuses the history of perception and instead orchestrates a future of nonpresence.

The antithesis of Afrofuturist diffusion (there being no bass response in the speakers), HyperSonic Sound technology represents the dystopian reverberation of Western science, dealing as it does in severance, detachment, and rational mutation. Whereas the internet transmits the bifurcations and minutiae of informational vicissitudes from a distance and across distance, the HSS pitchshifts telepistemology's mandate into a paranoid echo of 'videodromotic transmission'. It does not take a great leap of association to suggest that the arcane dark signal in the science-fiction film *Videodrome*[10]—causing neural transfiguration and hallucination—has technologically evolved from 1983's cultivated science-fictive blip into a martially distributed channel of mental destabilisation and spatial dislocation by 2019.

The development of such technologies suggests that we have broached Philip K. Dick's projected future in which covert transmissions are localised by neural markers.[11] And it is precisely this dynamic of cognitive rupture that sits between the cross hairs of the HSS's heterodyned emission. In an act of acoustic double-cross, the HyperSonic Sound System quietly mobilises its position as a whispering parasite. Whilst its viral objective is the transmogrification and multiplication of the inner voice, its aim is to awaken a Siamese consciousness. Orchestrating this unsound economy of (ir)rationality is a system that invests in surfaces so that it can trade in the depths that comprise the undead currency of the self.

9.   A sign—such as a subject declaring that she is hearing voices—that communicates an individual's declining mental health to a wider population.

10.   *Videodrome*, dir. D. Cronenberg (1983).

11.   P.K. Dick, *Radio Free Albemuth* (Westminster, MD: Arbor House, 1985).

THE LAMENT

Eleni Ikoniadou

Yet meet we shall, and part, and meet again,
Where dead men meet, on lips of living men.
— Samuel Butler

Chorus Leader: What sweet relief to sufferers it is to weep, to mourn, lament,
and chant the dirge that tells of grief!
— Euripides

All over the ancient and modern world, death is a woman's business. Women wash, dress and decorate the corpse, and then sing it to its final resting place with a lament. Lamentation is an extreme expression of sorrow that precedes every other form of oral ritual, and has led to the creation of the oldest epic poems across human culture.

A traditional burial includes the wake, the procession to the cemetery, the funeral itself, and memorials at a future point in time. All four might be accompanied by one or more female lamenters. These are usually older women dressed head to toe in black, vocalising the horror of the loss with a chilling lament, ranging from talking and singing to sobbing, keening, and wailing, for the one no longer there.

Keeners can be friends or relatives of the deceased and their family, part of the wider community, or hired professionals, briefed about the dead person's character, background, and history prior to the funeral. The ritual may include these participants lifting their arms in the air, clapping their hands in unison, beating their heads and breasts, and pulling their hair. There is a rhythm to the performance, though performing here doesn't mean faking it, as even hired lamenters are emotionally invested in the particular deceased person they are lamenting.

The process starts with a light wailing during the wake, which lifts to a crescendo during the procession, subsiding briefly in the course of the church service only to rise again on the way to the grave, and soar to a climax during the lowering of the coffin into the ground. But this unspoken rhythmic rule is one of few repetitive elements, as the lament is almost always improvised.

This is known as the primary lament, erupting spontaneously from overwhelming grief, barely controlled by the lamenter. The sound of death is a formidable force, taking over the vocal cords of the woman and gushing out of her mouth like a torrent of wildly manifold configurations: sophisticated literary content gives way to street language, poetry turns to swearing, and stormy outbursts are preceded by calm seas in the voice and intonation of the mourner.

At its core, the dirge is uncontrollable and unknowable, making it impossible to repeat or own entirely. This explains how the same mourner can produce elegies of entirely different form, style, and quality. However, it would be wrong to assume that laments merely derive from within. While the main lamenter bursts out improvised words and sounds, the surrounding women incessantly feed her with information about the departed, which she effortlessly incorporates in her keening in real time. Therefore, the lamenter, in addition to speaking, is also always listening.

And yet a lament typically contains more than just pure facts about the history of the dead. Entangled with it is information apparently known only to the lamenter, and which she seems to have garnered from unknown sources. The lamenter criss-crosses the deluge of information she holds about the deceased's past with the data received in real time, adding speculative material and processing it all at incredible speeds while vocalising it. In so doing, she is making things once considered private into a part of the permanent record.

At its climax, the lamenting voice leaves the past and present of this world to open up a door to the otherworldly. More than sonically expressing the grief and pain of loss, lamentation is rooted in a concrete ancient belief in the afterlife. Accordingly, wakes and burials are of great significance, as the last chance to prepare and equip the deceased for their journey to the underworld. Some of their favourite things—coins for the boat-fare to the other side, praise for their lives, messages to pass on to other dead, are gathered by the lamenter herself.

Lamenters are actively interested in the dead body's fate in the next world, and are seen as being capable of opening up channels of transmission between the living and the dead through their unsettling vocalisations. The main lamenter is not to be interrupted at any cost during the build-up of a lament, and is typically feared, admired, respected, but also mocked and hated, largely by the men, who hold lamenting to be dangerous witchcraft. The threat that moirologists are perceived to pose to the social order owes to the extreme uncertainty that such an orgiastic state of grief carries with it. But it is also to do with the fact that, in lamentation, women were allowed an isolated moment of speaking out. Hence lamenters would often deviate from the particular death they were mourning and move on to other sensitive, political, private, and public matters, commonly untouchable by females, and in some cases even by males. Mostly, however, the terror of the lament lies in its extemporaneous, untamed, inhuman dimension; that which reveals it as a sonorous force of unspecified destiny and unknown origin, separate from the body that hosts it.

In the lament, we find an urgency to channel the alien, all-devouring unseen that lies beyond this world. The lamenter becomes a transducer of death into sound, an acoustic passage between different orders of the real, devising a direct encounter between incompatible realms. Her unearthly incantation leading across, transferring to or from, vocally mediating and negotiating the ceaseless trade between the living and the dead is primordial, pre-mammalian, both ancient and yet to come.

UVO

UNIDENTIFIED VIBRATORY OBJECTS

THE JODHPUR BOOM

Paul Purgas

Reported to have taken place in central Rajasthan on the morning of Monday 12 December 2012, the Jodhpur Boom remains an unexplained sonic phenomenon experienced by the inhabitants of this desert region of northern India. Occurring at 11:25am, a deafening sound was released from the sky, creating a shock wave that shook the streets of the town of Jodhpur, causing widespread fear and panic. Initially believed to have been a jet aircraft or an ammunitions explosion, this was soon ruled out since no known flight path passed near the town and there were no visible signs of damage that might correspond with the use of explosives. An investigation by the Indian military discounted the possibility of an aircraft being responsible, owing to an embargo on flights over populated areas and there having been no possible cause of an explosion in the region. It also proved difficult to explain the deafening amplitude of the boom itself and its corresponding shock wave, which was reported to have been significantly louder than the effect of a jet aircraft crossing the sound barrier. To further add to the mystery, it transpires that throughout December 2012, a month linked to various eschatological beliefs of global cataclysm and Mayan apocalypse, numerous other booms were also heard at sites across the planet.

CBS News reported that on 4 December 2012 residents in several communities in central Arizona reported a similar deafening boom. The United States Geological Survey reported no significant earthquakes had taken place that could explain the sound, and local police were unable to find a cause. In Warwick, Rhode Island, the police department received almost one hundred phone calls relating to a loud noise that sounded like an explosion on the evening of Monday 3 December 2012. Residents in nearby Narragansett Bay also reported a low droning noise coming from the water that began about an hour or so after the initial boom was heard. Following investigations by local authorities, they were equally unable to identify the source of the boom or, on this occasion, its accompanying eerie drone. Alongside these, similar events were also reported to have taken place in Georgia and Texas at around the same time.

One of the explanations for these occurrences has been the possibility they may have been extreme examples of a skyquake, a phenomenon described as inordinately loud thunder that manifests even though there are no clouds in the sky large enough to generate lightning. Among those with military experience, the sound of the skyquake is most often likened to the boom of cannon fire. Recurring sites are often named to this effect, such as the Barisal Guns near the Bay of Bengal in Bangladesh, Hanley's Guns in Victoria, Australia, and the Guns of Seneca encountered around Seneca Lake in New York State. Many early settlers near the Seneca area were told by the Haudenosaunee Iroquois people that the booms were the sound of the Great Spirit continuing his work of shaping the earth, and later the phenomenon entered into popular culture through the 1850 short story *The Lake Guns* by James Fenimore Cooper, author of *The Last of the Mohicans*.

Current theories for the skyquake range from the possible release of volatile gas from deep underground deposits, seismic vibrations and minor earthquakes, through to more recent theories of cosmic radiation or the possibility that the booms might perhaps be the sound of distant thunder focussed anomalously as it travels through the upper atmosphere. Whatever the definitive answer may be, the fierce alarm of the skyquake is unquestionable in its force and scale, reaching deep into the primordial psyche, exposing and rupturing the daily cycle of life and revealing humanity's futile attempts at dominion over Nature. To the inhabitants of Jodhpur, the deafening boom that shook the town may have resembled a holy story from the *Brihadaranyaka Upanishad,* one of the oldest Hindu scriptures. Here, Lord Brahma the creator speaks to his three children, the gods, demons and men that inhabit the earth. His message, delivered as an overwhelming burst of thunder, instructs his children to obey the cardinal virtues of compassion, charity, and self control, with the booming voice of Brahma manifesting through one deafening single syllable: 'DA.'

It is perhaps with this more otherworldly reading that the events of December 2012 and their underwritten Mayan script map out an alternatively discernable series of events: that these volatile blasts might emanate from a source unknown to us, as emissions or even as a voice from beyond in a language so alien as to be overwhelming. Utterances of warning or even welcome that appear so violently manifest as to simply evoke terror to the human senses. As yet their interpretation remains a mystery that may only be uncovered when these booms return en masse to echo across the planet. Let us hope that, when this day comes, their message is revealed to us as one of mercy.

THE BLOOP

S. Ayesha Hameed

In 1997 Drexciya, an electronic band from Detroit, released the album *The Quest*. Its liner notes tell the following story:

> During the greatest Holocaust the world has ever known, pregnant America-bound African slaves were thrown overboard by the thousands during labour for being sick and disruptive cargo. Is it possible that they could have given birth at sea to babies that never needed air? Are Drexciyans water-breathing aquatically mutated descendants of those unfortunate victims of human greed? Recent experiments have shown a premature human infant saved from certain death by breathing liquid oxygen through its underdeveloped lungs.[1]

This story draws on a particular practice in the history of transatlantic slavery, the jettison of slaves for insurance purposes. Here Drexciya posit an alternate ending: that the foetuses of the pregnant women thrown overboard adapted from living in amniotic fluid to living underwater. These newly adapted underwater people and their descendants set up a Black Atlantis called Drexciya at the bottom of the ocean. The covers of subsequent albums trace the evolution of Drexciyan creatures from gill-breathing aquatic creatures, to flippered-feet wave jumpers, to outer space explorers. These later albums study the pragmatic infrastructures under which Drexciya and other underwater cities functioned, their ecologies, etc.[2]

Also in 1997, the National Oceanic and Atmospheric Administration (NOAA) discovered a very powerful ultra low frequency sound coming from the ocean surrounding

---

1. *The Quest* (Detroit: Submerge Records, 1997).
2. See for example *Neptune's Lair* (Germany: Tresor Records, 1999).

the southern tip of South America, and named it 'the bloop'.[3] Picked up by autonomous hydrophone arrays, the sound was detectable from over five thousand kilometres away. Its audio profile resembled that of a living creature, but was exponentially louder than sounds produced by the loudest animal, the blue whale. At the time, the NOAA's Dr Christopher Fox noted that the signature of the sound varied rapidly in frequency, a trait characteristic of marine animals. This begged the tantalising question: Was there a creature even bigger than a blue whale lurking at the bottom of the sea?

The NOAA eventually decided that the sound came from a large ice quake at the bottom of the ocean. The confusion surrounding the bloop makes for another reading of life under the sea though. At a sonic level, the difference between sentient life and the environment became blurred and unreadable. Ice could make the sound of an underwater animal. Perhaps this confusion provides us with a way of making sense of Drexciya's experiment—to blend the human with the nonhuman as a form of adaptation and survival. Their experiment blurs the parameters of the ecological to make for another possibility for life—which resonates with the blurring of the sonic threshold performed by the bloop.

In 2008 M NourbeSe Philip published *Zong!*, a cycle of poems written to mark the massacre of slaves thrown from the slave ship Zong in 1781.[4] All the poems draw solely from the text of the two-page document surviving the Gregson v. Gilbert court case, the insurance claim made in court after the jettison.[5] In the first poem the word 'water' (with a few others) is splintered and fractured along the length of the page; and under a line in the gutter of the page at the bottom is a list of imagined names of the slaves thrown overboard. The scattered words read as the depth of the sea from surface to ocean floor, the line drawn across the bottom of the page. In her performance of these poems Philip pulls the words to their breaking point, the completion of each utterance of the syllables in the word 'water' catching at her throat. It sounds as if she is drowning. In pronouncing the word 'water', water becomes both the subject and the object of drowning. It is painful and violent.[6]

3.   D. Wolman, 'Calls from the Deep', *New Scientist*, 15 June 2002, <http://web.archive.org/web/20130106141048/http://www.science.org.au/nova/newscientist/102ns_001.htm>.

4.   M. NourbeSe Philip, *Zong!* (Middletown CT: Wesleyan University Press, 2008).

5.   O. Berrada, 'Defend the Dead: Omar Berrada on M. NourbeSe Philip's Zong!', Radio Broadcast, Chumurenga/Pan African Space Station, 2015, < https://www.mixcloud.com/chimurenga/defend-the-dead-omar-berrada-zong/>.

6.   Ibid.

The sound of suffocated drowning that infuses Philip's performance of *Zong!* and particularly the word 'water' turns the voice into something not human, almost watery, environmental. It crosses the threshold between the human and the underwater environment. The sound of the word 'water' and its meaning become as liminal an entity as the bloop, but this time they cross the threshold in the opposite direction, where the human voice becomes something inhuman, watery, monstrous.

Both the bloop and the sound of the word 'water' blur their human and inhuman qualities, and this indeterminacy is what fuels the speculation of an underwater Drexciya. But there is something monstrous in the uncanniness of both. The bloop evokes the murky horror of imagining a screeching creature in uncharted depths that is more gargantuan than a blue whale. With the performance of the word 'water' there is the horror of hearing a voice stretching beyond human detection towards its own annihilation. This is a measure of the horror embedded in Drexciya's imagining of the possibility of life growing in the bodies of drowned enslaved women thrown overboard.

It raises the question: To what extent is horror a fuel for resistance?

And: To what extent is adaptation a form of agency?

It casts the sonic as a vibrating vehicle transmitting this knife-edged form of survival, and opens the discourse of adaptation/survival to nonhuman forms of life. Part of its power lies in looking directly at the moment of simultaneous horror and annihilation, into its subaquatic Cthulhu-esque face. And to make the reverberations of that horror into progenitors of a form of response and survival whose indeterminacy finds echoes in the bloop.

> ...but some day the piecing together of dissociated knowledge will open up such terrifying vistas of reality, and of our frightful position therein, that we shall either go mad from the revelation or flee from the deadly light into the peace and safety of a new dark age.[7]

7.   H.P. Lovecraft, 'The Call of Cthulhu', in *The Best of H.P. Lovecraft: Bloodcurdling Tales of Horror and the Macabre* (New York: Ballantine Books, 1982), 76.

SOUND OF THE ABYSS

Eugene Thacker

> Music is the last enunciation of the universe.
> — E.M. Cioran

Sub-Bass from Deep Space

In January of 2009, scientists at NASA's Goddard Space Flight Center announced receiving transmissions of an unexpected and unexplained cosmic sound. The NASA team's huge, balloon-like satellite, which is immersed in approximately 500 gallons of ultra-cold liquid helium, was originally launched in 2006, ascending some 120,000 feet into the atmosphere, where it was to detect subtle heat emissions from very early star formations. Instead, it became a receptacle for a kind of cosmic sub-bass. As one of the scientists noted, 'instead of the faint signal we hoped to find, here was this booming noise six times louder than anyone had predicted'. A NASA press release noted that 'detailed analysis ruled out an origin from primordial stars or from known radio sources, including gas in the outermost halo of our own galaxy. The source of this cosmic radio background remains a mystery'.

While radio emissions from space are not uncommon, what makes these sounds a mystery is not just their magnitude, but an apparent lack of point of origin. In short, there do not seem to be enough galaxies around to possibly produce a sound of such magnitude. Neither supernovas, aliens, or the Death Star are capable of generating such a sound. Perhaps we were witnessing what alchemist Robert Fludd once described as the 'celestial monochord'.

In the early seventeenth century, Robert Fludd produced a diagram linking the Ptolemaic universe to musical intervals, providing a means not only of viewing the cosmos as sonically ordered (a principle of sonic reason, as it were), but of comprehending the possibility of a divine superchord that would be responsible for all other sounds.

While the NASA report makes no such occult claims, it is nonetheless interesting because it hints at a theme that is, I think, at the centre of extreme music genres today, namely the relationship between *sound* and *negation*. Now, we commonly think of the relation between sound and negation I terms of the negation *of* sound. And this in turn, relies on the well-worn dichotomy of sound and silence.

One can explore all sorts of combinations within these relationships. But that is not my aim here. What I'd like to suggest is that, when thinking about sound and negation, negation is often understood as something that happens *to* sound, or, alternately, that happens *in* music. By contrast, we can take a different approach and ask: Can music or sound itself be a negation? That is, is there a negation that is not something that one does *to* sound or music, and that is also not something that simply *produces* silence?

In a way, this is what black metal genres do, presenting us with forms of negation that are co-extensive with music and sound. For instance, old school Norwegian Black Metal, with its intentional use of lo-fi recording equipment and stripped-down song structures, presents a music in which the separation between individual instruments ends up in decay and indistinction, a melody that melds into anti-melody. The 'necro' sound presents music as a negation of the soundness of sound, an encrusted, distorted music that is about to rot not only the musical foundation of melody, but the physical substrate of music itself. Similarly, Doom Metal and Funeral Doom Metal take the most basic element of music—its temporal flow and its existing in time—and negate it by making music that is grave, *gravitas*, leaden and weighted down with the gravity of melancology and pessimism. Doom Metal presents music of a tempo so *grave* that it negates tempo. Finally, Drone Metal, with its minimalist dissipation of all music into a monolithic, dense line of sound, presents us with the whittling away of all harmony into a single, thick, absolute tone, collapsing the musical spectrum into a dense black hole. In each of these examples, Black Metal presents a music that negates some aspect of musical form—a song against all melody, a rhythm against all tempo, and a harmony against all tonality. What results is not an absence of music per se, but rather a form of anti-music, expressed through music. At its limit, Black Metal brings us back to an even more basic distinction—that between music and sound, with the former continually threatening (or promising) to dissolve into the latter.

Let us return briefly to the NASA report of cosmic sound. One idea evoked in the report is the notion of a sound without a point of origin. We know that sound is a physical phenomenon. The basic physics of acoustics necessitates an origin of sound production—say, the vocal cords, or a woofer in a speaker—that produces sound waves which then radiate through the air. Such waves physically move us, brush up against us, and pass through us, a portion of them being registered by our ears—but also, on occasion, in our chest, or in our breath. In philosophical terms, the physics of sound production looks very much like a Neoplatonic emanation of immaterial forms. This is a Plotinean sound, the sound of radiation, emanation, the sound that is outpouring and outflowing.

But what the NASA report seems to indicate is that there is a sound that has such magnitude, such density, that there is no point of origin that can contain it, much less produce it. Taking some liberties here, we might say that the sound is so much sound, so much in excess of itself, that it is a sound that paradoxically has never been produced. This is a Kantian sound, a sound that is dynamically sublime in relation to its point of origin. This sound exceeds itself and thus eclipses its own point of origin. The result is an enigmatic sound that is so much sound that it negates itself, becoming...What? Silence? Noise? Or something else altogether?

And from here we move yet another step, to a third type of sound. As one of the NASA scientists notes, in order to have enough galaxies in the universe to produce this sound, 'you'd have to pack them into the universe like sardines [...] There wouldn't be any space left between one galaxy and the next.' Now, aside from this rather Lovecraftian image of cosmic sardines sonically descending upon the Earth, what is interesting here is the notion of a sound without any point of origin, a non-directional sound. This is different from the Neoplatonic sound (a sound that radiates from a point of origin), and different again from the Kantian sound (a sound that exceeds and eclipses itself). This is the sound of Schopenhauer, a Schopenhauerian sound. This sound does not have an origin to negate, because there is no origin to negate. But it is also not simply a positive sound, a fecund sound that continuously pours itself forth, continuous sound as a gift of the heavens. Rather it is a sound that is, at the same time, pure nothingness, a presence that only asserts its absence.

THE HUM

Kristen Gallerneaux

One day as I sat at the breakfast table, skimming through a pile of newspapers that should have been recycled months before, I felt the vague disturbance of familiarity. As the smoke cleared, I realized that the photograph on the front page of *The Detroit Free Press* was of my old apartment in Windsor, Ontario. Apparently the sickly yellow stucco walkup in Sandwich Town had become newsworthy as a victim of 'The Windsor Hum'. Reports of Hum-like activity are global, reaching back into the 1830s, and find a push-pin presence spreading over maps throughout the 1970s. Wherever it appears, etymology melds with geolocation: the Taos Hum, the Bristol Hum, the Auckland Hum. In early 2011, Windsor developed its very own Hum, a mysterious infrasonic event spiking deep at 35Hz. When residents woke late one night to a low-frequency rumble, slashing open their curtains to yell at an idling car, booming bass—they found empty, dark streets.

In 2002, while standing insomniac-prone in that same Windsor walkup, I looked out the kitchen window to find the sky on fire. A total apocalyptic vision over Detroit. Gigantic orange plumes trailing up into a gradient of anemic ochre, wretched green, and hazy purple—a low-grade thrum prickling at the soles of my naked feet. Assuming some great industrial disaster was about to roll toxic fumes over the river, I pounded on my roommate's door and stuttered my worries about the cataclysm in the sky. This is when I first heard the (pillow-groaned) words: 'Don't worry. It's just Zzzzzug Island.' Zug Island will play itself out in a moment.

Describe the Windsor Hum. A deep time bass rattle; a quivering in the gut. The creak of double-glazed windows with an angry bee caught between two planes. Night terrors. Not everyone can 'hear' the Hum, but the vibroacoustic effects of infrasound—sound that exists below the range of human hearing—can cause suffering, instigating fatigue, insomnia, depression, anxiety, and migraines. Most victims of the Hum describe it as something *felt* more than *heard*, as their bootlegged bodies suffer incessant mono- tone pressure beating on their eardrums. Rational finger-pointing towards local heavy industry was counterbalanced with viral conspiracies: trending UFO reports, ionospheric HAARP interventions, and flyovers by experimental military aircraft.

The Hum and infrasound alike can mimic the tropes of a traditional haunting. In the early 1980s, scientist Vic Tandy was likely surprised to find himself collaborating with psychical researchers, tracing the cause of a recent 'haunting' outbreak in his laboratory to the installation of a ventilation fan. That recent cold-sweat feeling of dread and the shadowy apparitions stuck in the corner of Tandy's eye were linked to inaudible infrasound being produced by the fan, a steady 18.9Hz. Tandy's vibrating eye- balls were allayed by the removal of the fan.[1] The Hum is a more holistic environmental phenomena, running counter to easy solutions. In its most-close reality, The Hum is a fake-out haunting—a physiological response to the invisible effects of the discord between environment and industry—an ominous protest of the Anthropocene in the form of infrasonic terror. Salt and steel dancing on air, down into the lungs.

There are other speculative celestial events that compete with the same affecting clash of The Hum. When the Tunguska Event occurred in 1908, the pressure of its explosive power jolted the needles on barographs around the world. Decades later, atmospheric researchers visiting the forest levelled by the Tunguska shock wave discov- ered that locals were reluctant to talk about the event. Homegrown folklore had been etched into place, to explain the explosion as a curse from Ogdy—the god of thunder and infra-bass—who smashed the forests and chased off the animals as a punishment.

Shifting now from a boom to a tinny crackle: Arctic explorers travelling into the Far North have reported experiencing sonified light displays in the sky, courtesy of the Aurora Borealis. Described by the poet Robert Service as rolling 'with a soundless sound, like softly bruised silk',[2] the wavering Northern Lights sometimes play samples of radio static, clapping hands, and the gnashing teeth of the *kiguruyat* spirits. Most common among Algonquian and Inuit tribes is the belief that this skyborne soundtrack is caused by spirits playing football with a walrus or a child's skull, or that the dead are trying to pass messages to the living. Engaging in an exchange with the sky has

1. See V. Tandy and T.R. Lawrence, 'The Ghost in the Machine', *Journal of the Society for Psychical Research* 62:851 (April 1998): 360–64.

2. R.W. Service, *Best Tales of the Yukon* (Philadelphia, PA: Running Press, 1983), 43.

its consequences: when the auroral spirits whistle to the living, 'they should always be answered in a whispering voice',[3] according to Arctic anthropologist Ernest Hawkes. The reality of aurora-produced sound has largely been shrugged off as an auditory illusion, placing the phenomena in the same contentious territory as The Hum. Current research seems to have settled on the explanation of 'electrophonic transduction', which is to say that the low-frequency VLF radio waves produced by aurora can turn long, thin objects—such as blades of grass, wire, and human hair—into antennae, vibrating signals into audible sound.

Coming back to the earthbound resonant mysteries of The Windsor Hum, attempts to trace its location (never mind its cause) have been an exercise in frustration. Like describing neural pain or ghost limbs, pinpointing where one flesh ends and the other ghost twin begins, the Hum's oppression seems to come from everywhere and nowhere. First, the semi-trucks idling on the crumbling Ambassador Bridge that joins Windsor and Detroit were blamed. But this theory belly-flopped into the river below.

Next, the salt mines that form a handshake between countries 1200 feet *underneath* the Detroit River were accused. Down there, a 1500-acre crystalline rock salt city makes a maze of itself—over a hundred miles of subterranean mining road loops itself into knots. Salt chamber walls are sheared off with explosives, crushed, and conveyor-driven up for Michigan's winter roads. The mine blasts decouple the slow capital production of the earth—but these operations were proven innocent, because they were inactive during the peak hours of the Windsor Hum.

It wasn't the salt or the bridge that were to blame, but heavy industry playing itself out as a slapback echo. In 2013, hard-nosed scientists finally captured the 'temporal and spectral' signature of Windsor's Hum, describing the process as being 'like chasing a ghost'.[4] Accusing fingers pointing towards Zug Island transformed into tenuous high-fives: the electric arc blast furnace at the US Steel plant was haemorrhaging infrasound and VLF waves across the border. These waves have been identified as the 'likely' cause of The Windsor Hum—and are the same waves believed to cause the elusive soundtrack of the Aurora Borealis. On Zug Island, the resistance of steel being magnetized back into its base elements reverberates like the wailing of entangled souls, while offsite it bleeds over the border, damped down into a low-pressure menace.

3.   E.W. Hawkes, *The Labrador Eskimo*, (Ottawa, 1916; reprinted New York: Johnson Reprint Corp., 1970), 137.

4.   C. Novak, 'Summary of the Windsor Hum Study Results', Global Affairs Canada, Government of Canada, 23 May 2014, <http://www.international.gc.ca>.

The Hum continues to beat the ears of the city in unpredictable fits of biomechanical violence; the noise is always there.

ALLIGATORS OF YOUR MIND

Dave Tompkins

The ashtray levitated with an alligator glued to its rim. Hijacked by the subconscious, the souvenir flew, passing over the cowbell and the other less gifted alligator ashtrays, traveling through psi wave turbulence, bound for a memory of Cuba, only to crash to the floor of a warehouse in Miami. Flight time was brief but left an impression. Novelty item and reptile became Event 176, joining the Florida registry of psychokinetically traumatized objects.

In January of 1967, German parapsychologist William Roll was called in to investigate a souvenir wholesaler in north-east Miami, located in the future Little Haiti. The merchandise at Tropication Arts, Inc. had been subject to outbursts of inertial rejection. A box of backscratchers grew wings. Zombie glasses shattered, beer mugs scootched, rubber daggers jumped shelf. Aisle clean-ups were on the rise, as other items took to the air. Faux leathers, a water globe, a spoon rest, plastic TVs that sharpened pencils, coconut heads, an artificial orange impaled on a cocktail pin.

Vacation doodads had become psychic artifacts. 'Beer mugs and zombie glasses being especially active', observed Roll in the *Journal of Parapsychology* in 1973.[1] Roll had been invited to Miami by Susy Smith, a parapsychologist who would later devise codes for peers to track her consciousness after death. For further insight, Tropication owner Arnie Laubheim summoned the 'Magic Chatterbox', an cynical illusionist known for gimmicks like the 'Baffling Bra' and the 'Whistling Bellybutton'. Also at the scene: an airline pilot, a Baptist minister, an artist who painted flamingos onto plastic purses, a substantial amount of dry ice, a German shepherd with a low tolerance for sudden

---

1.   Refers to a bad mix of Tiki spirits whose resulting hangover is likened to a shot in the head, a common practice in on-screen zombie resolution.

movements by the otherwise inanimate, an ice dancer, and a police officer ridiculed by his co-workers for being there.

Some Tropication employees blamed the disturbances on the ghost of Laubheim's recently deceased pet squirrel monkey. Low-flying jets and 'atmospheric vibrations' were ruled out, but geomagnetic perturbations remained in play.[2] Roll and his colleague Gaither Pratt itemized each disruption and committed them to a map, a vectorized scrum of things going where they shouldn't. They also tried baiting the poltergeist, a path of scientific inquiry that led to Julio Vasquez, a disgruntled shipping clerk who'd immigrated to Cuba with his mother in the sixties. Designated as the psychokinetic agent, the nineteen-year-old was sent to the Psychical Research Center in Durham for tests, including a brainwave review and word association games triggered by the disturbed merchandise. If Vasquez's subconscious was at the controls, the normal restrictions of gravity, to say nothing of reality, did not apply. The kid went along with it, giving the Unexplained a shrug. He joked that the ghost was merely tired of the amber beer mugs, which were last seen 'moving in a northwesterly direction'.[3]

Vasquez was also under stress from living with his mother (see *Carrie*)[4] and felt that Laubheim was as phony as the goods he trafficked, not unusual in a state with a rich history of hoaxing. He may have been telekinetically venting against his boss, who'd once accused him of arson, or Tropication's carefree customers, tourists free to return home with their tans and *tchotchkes*. Recurrent spontaneous psychokinesis (RSPK) can be triggered by relocation, in this case an airlift from Cuba to Miami, events and movements beyond Vasquez's control, as if the flying objects were projections of displacement attributed to Cold War *exilio* stress.

Bill Joines, an Electrical and Computer Engineering professor at Duke University, collaborated with Bill Roll on the Miami case. A former missile radar engineer at Bell Labs, Joines could calculate the trajectory of a warhead launched from Alpha Battery, a covert Project Nike base in the Everglades.[5] He could also take radiation measurements

---

2.   W.G. Roll and G. Pratt, 'The Miami Disturbances', *Journal of the American Society for Psychical Research* 65 (1971).

3.   W.G. Roll, D.S. Burdick, and W. Joines, 'Radial and Tangential Forces in the Miami Poltergeist', *Journal of the American Society for Psychical Research* 67 (1973).

4.   Carrie's telekinetic outbursts rebelled against an overbearing, abusive mother, as well as high school bullies. Flying fire hoses and cutlery proved to be an effective defense, as evidenced in the landmark case *Carrie v. High School Senior Prom*. Also see: Briteway Billy, a supermarket gopher mascot that became the spirit animal of a psychokinetic clerk who pummeled her boss to death with non-perishables. From the minds of Stephen King and British screenwriter Nigel Kneale, respectively.

5.   The Hercules Nike site was one of four missile sites constructed in South Florida after the Cuban Missile Crisis

of one's personal psi field. Under constant surveillance, Vasquez was treated as a signal, analysed for attenuation and decay curves. Perhaps he was a transmitter, his broadcast pirated by his own subconscious and propagated over psi frequencies, interference picked up by one of the many rogue radio operations in spy-riddled Miami at the time. Location is everything, even in the metetherial vortex. 'The metetherial environment is equivalent to the "subliminal self", where the self extends beyond the borders of the familiar', wrote Roll in the *American Journal of Parapsychology*. To Julio Vasquez, the borders of the familiar—his birthplace ninety miles south—were contracting and expanding at once.

According to Gaither Pratt, the forces were selective. He would record himself doing a running account of all activities at Tropication, 'in the manner of a sports announcer', as if doing his best Marv Albert while the psychokinetic agent whizzed down the aisle in a go-kart full of backscratchers and shot glasses.[6] I imagine Gaither capturing his own golf whisper when a part-time medium appeared on site to perform an exorcism, in hopes of rescuing inventory from further obliteration. ('The Thing was giving the business to the business', wrote Susy Smith.)[7] Using ferns, cacti and incense, the medium constructed an altar for one of the toy rubber gators after claiming to have witnessed a full-scale alligator 'spirit entity', possibly prehistoric, hovering in one of the aisles.[8]

Hurricanes aside, the last time an alligator caught air in Florida was during a botched robbery attempt in Palm Beach, when the perpetrators tossed a live three-footer into the drive-thru window of a Wendy's. Though Julio Vasquez shared an emotional resonance with the alligator ashtray, information stored in the object itself was privy only to his subconscious. In RSPK terms, both object and agent were systems, in this case, tropic subsystems exchanging personal information and secret histories. Of course, information on real alligators is accessible to all from a respectful distance, having introduced themselves into the Florida golfing population to restore the ecologic balance. Their trunks low to the earth, these reptilian mud-bathers are a Miami sub-frequency, equipped with extremely good hearing and dome pressure receptors in their snouts, which are sensitive to vibrations and signal waves. All with a stone-cold stillness while

---

in 1962. The missile barns are still intact and its ghosts are not supernatural. With the mass influx of asylum seekers from Cuba and Haiti in 1980, D Battery was repurposed as the Krome Service Processing Center, a model for America's current immigration prison policy. Krome is now overseen by ICE.

6.   Roll and Pratt, Ibid.

7.   S. Smith, *Prominent American Ghosts* (New York: World Publishing Company, 1967.)

8.   W.G. Roll, *The Poltergeist* (New York: Paraview, 1972.)

gnats jitter in their nostrils. In his book *Miami: City of the Future*, T.D. Allman drives to what he believes is the end of Miami, only to find an alligator lounging in the middle of the road, a nap cordoned off by traffic cones. The alligator is listening to

a distant roar, a faint rumble, a little like breaking waves, a lot more like the hum of a freeway. The alligator is listening to the sound of quicksand being metamorphosed into concrete, of swamp and scrubland transforming itself, almost overnight...the alligator is listening to Miami.[9]

Florida has always been tuned into caiman frequencies, whether poaching for boots and baggage,[10] or hearing the legend of Uncle Monday, a Hoodoo root doctor who escaped a plantation in Georgia and fought alongside the Seminole resistance during a siege at Fort Maitland in Orlando. Monday avoided recapture by transforming into an alligator and vanishing into the swamp, amid a chorus of low-end gator bellows.[11] Recognized for their long-range acoustic signaling, alligators are also infrasonic agents, vibrating their spines at 10 Hz, producing bass waves in drainage canals and ponds, making the water dance for territory and interested partners.[12]

Much of what transpired at Tropication's altar of the alligator was 'percussive'.[13] Susy Smith indicated that the 'bric-a-brac boogie' happening in the aisles was a symptom of the poltergeist reveling in the sound of itself.[14] In the mid-1980s, the shattered glass sound that frequented Miami radio could be traced to Music Specialist Studios in Little Haiti, where producer/audio engineer Pretty Tony Butler recorded himself smashing champagne glasses to make teenagers dance at skate rinks. Violent glass dispersal would be a signature effect in Butler's laser-clean electro sound. Pretty Tony was known to report to work dressed like an eccentric World War I pilot, wearing aviator flaps,

9.   T.D. Allman, *Miami: City of the Future* (New York: Atlantic Monthly Press, 1987).

10.   In *Swamplife* (Minneapolis: Minnesota University Press, 2011), anthropologist Laura Ogden writes about the 'refrain of the flesh' and 'becoming alligator' when hunting and skinning, and how the 'territorial practices of humans and nonhumans entangle and reshape each other'.

11.   'Slide We Fly' said Kool Keith, a man who once claimed to be half-alligator himself. The Uncle Monday legend comes from Zora Neale Hurston's collected oral histories of South Florida.

12.   Writing about gator signaling in *Copeia*, Vladimir Dinets refers to 'slaps', a jaw-to-surface 'advertising call' conducted in swamps, canals, and zoos (V. Dinets, 'Long-Distance Signaling in Crocodylia', *Copeia* 2 (September 2013), 517–26. See also: 'Slaps', a subgenre of Bay Area hip hop popularized by E-40, as well as Mac Dre, an Oakland rapper known to have 'gator-back throat' from smoking too many Backwoods. Please see also: 'Dredio'.

13.   W.G. Roll and W.T. Joines, 'RSPK and Consciousness', *Journal of Parapsychology* 2 (1977).

14.   Smith, ibid.

anti-bug shear goggles, and a necklace of studio patch cords. A more common flying phenomena in Miami during the 80s was people skylining on cocaine—off the glass and up the nose, as narconomics fueled a downtown boom of mirrored, curtain-wall high rises. Nobody said 'The Explained' had to be rational. At the Music Specialist, there would be several reincarnations of hit acts like Debbie Deb and Freestyle, as well as a Bentley that appeared to start by itself. The car was haunted by the World Famous Sweetback, a Miami radio DJ whose voice had been programmed into the console to offer customized safety cues. *Close the fucking door.*[15]

The studio walls of Music Specialist were reinforced with sand poured between the layers of two concrete walls—vacated mollusc grit dredged from the ocean bottom and cemented into building materials. The gator that T.D. Allman found napping at the end of the road was a living fossil tuned into limestone tape loops, ghosts from shells disinterred from the Benthic Zone, embedded deep in Miami's aqueous psyche. These are the psi waves of a city built over a swamp and covered in skeletons,[16] whose subconscious—to say nothing of its acute racial and climate tensions—has often been suppressed to the point of denial.

The former Tropication space at Northeast 54th Street is a few doors down from Toute Divisions Botanica, seller of herbs, tinctures, powders, customized mojo bags to protect the home, and finely pulverized geology for transduction.[17] Like much of Little Haiti, this strip of businesses sits seven feet above sea level, prime real estate for developers who are speculating an inland oceanfront while trying to push out the neighborhood's longtime black Caribbean residents.[18]

The dozing alligator could've been listening to the sound of human displacement as much as freeway infrasonics and landscape in transition, the underwater memory of Miami's own future. A nightmare fully awake in flight, with souvenirs left behind. A water globe, a shell that heard it all before, a floating ashtray.

---

15.   Allen Johnston, Personal correspondence, 19 June 2016.

16.   J.E. Hoffmeister, *Land from the Sea: The Geologic Story of South Florida* (Miami: University of Miami Press, 1974).

17.   Soil samples from the Tropication space have been mixed with magnetite for a modular 'dirt synth' patch for low-frequency mineral transduction. Research and practice conducted by fellow contributor Kristen Gallerneaux.

18.   Florida historian Paul George wrote that the borders of Little Haiti are 'subjective', especially when serving the interests of developers. Part of the area was Lemon City, one of Miami's earliest African American and Afro-Caribbean settlements. The CLEO Institute is a Miami non-profit currently doing important work in climate activism and gentrification.

RESONANCE

Erik Davis

In 1971, Terence McKenna, his brother Dennis, and some friends travelled to the small Columbian village of La Chorrera in search of botanical wisdom. One evening, they settled into their hammocks after consuming a pile of fresh psilocybe mushrooms. As they began tripping, Dennis noticed an otherwise inaudible buzzing in his head. Terence asked him to imitate the noise, but Dennis demurred. Then, as Terence tells it,

> the drizzle lifted somewhat, and we could faintly hear the sound of a transistor radio being carried by someone who had chosen the let-up in the storm to make his or her way up the hill on a small path that passed a few feet from our hut. Our conversation stopped while we listened to the small radio sound as it drew near and then began to fade.[1]

What happened next was nothing less than a turn of events that would propel them into another world. For with the fading of the radio Dennis gave forth, for a few seconds, a very machine-like, loud, dry buzz, during which his body became stiff. After a moment's silence, he broke into a frightened series of excited questions. 'What happened?' and, most memorably, 'I don't want to become a giant insect!'

This blast of high weirdness unleashed a flood of bizarre ideas in Dennis, while giving the McKennas the core theoretical and expressive principle of the Experiment at La Chorrera they would subsequently perform: the principle of *resonance*.

In his book *Brotherhood of the Screaming Abyss*, Dennis provides a formative example. During high school band practice, his instructor plucked the pitch of A on a bass string, which caused nearby strings tuned to A to vibrate as well. Resonance here means two systems entering into an energetic relationship mediated by frequency,

1.   T. McKenna, *True Hallucinations* (San Francisco, CA: Harper San Francisco, 1993), 53.

a mutual oscillation that, once begun, allows the second string to continue to sound even if the first string is dampened.

The physical phenomenon of resonance operates in many different systems, among molecular particles, in neural tissue, and in a host of electronic technologies. It is one of the fundamental figurations of a cosmos that vibrates about as much as it does anything else. But resonance also resounds within symbolic, philosophical, and phenomenological frameworks. The term derives from *resonantia*, the Latin 'echo', and one thing the physical phenomenon echoes are magical doctrines of sympathy, such as the ancient correspondence between microcosm and macrocosm enshrined in the hermetic doctrine 'As above, so below'.

This essentially erotic model of the resonating cosmos becomes part of the modern magical underground as well as a significant topos in alternative medicine. Contemporary esoteric and New Age practitioners operate in a vibrating realm of 'energies' that manage to follow physical wave dynamics while eluding conventional measurement devices. As such, contemporary spiritual or esoteric discourses based on 'resonance', 'vibrations', and esoteric 'frequencies' are frequently discounted as pseudo-science. But sometimes, as with the McKennas, a zone of indeterminacy is discovered, where the systems that begin to resonate themselves cross multiple fields of physics, sound, symbol, and experience.

In his book *Reason and Resonance*, musicologist Veit Erlmann argues that even the *physical* phenomenon of resonance presents a challenge to the rationalist legacy of modern philosophy. With their ocular bias, rationalists characterize the mind as a kind of mirror capable of capturing accurate representations of the outside world while remaining fundamentally separate from that world. Resonance, on the other hand, is a phenomenon of *conjunction*, of the blurring of the boundary between subject and object. Rationalists ignore or suppress resonance, which nonetheless remains, in contrary traditions like Romanticism and phenomenology, 'inextricably woven into the warp and woof of modernity'.[2]

So let us listen again to the curious transistor radio that night in La Chorrera, the 'small radio sound' that catalysed Dennis's inner signal into wild expression.

In *Understanding Media*, McLuhan underscored the connection between the radio and resonance's alternative to rationality. 'The subliminal depths of radio are charged with the resonating echoes of tribal horns and antique drums,' he wrote in 1964. 'This

---

2.   V. Erlmann, *Reason and Resonance: A History of Modern Aurality* (New York: Zone Books, 2014), 15.

is inherent in the very nature of this medium, with its power to turn the psyche and society into a single echo chamber.'[3] Behind McLuhan's claustrophobic and colonialist language—with its hint of Jung's 'subliminal depths'—is the spectre of Hitler's radio performances, and the belief that the fascist ability to mobilize such irrational and seemingly 'mythic' identifications on the part of the crowd was directly tied to the medium of radio.

But McLuhan's analysis was not only symbolic but also formal, since the phenomenon of resonance also defines the technological action of radio tuners: in order to select and amplify a single radio frequency out of the thousands picked up by an antenna, radios use an adjustable oscillating circuit, known as a resonator, to resound with the desired frequency.

In a diary entry, Dennis described the sound he heard inside his head as 'something like chimes at first, but gradually becoming amplified into a snapping, popping, gurgling, cracking electrical sound'. Such sounds appear not infrequently in anecdotal accounts of psychedelic experience, especially in response to high doses of tryptamines like psilocybin and DMT. By attempting to give physical voice to this virtual or 'inner' sound, Dennis responded to the radio's resonator by probing the resonating capacities of the various cavities in his body in order to find, and construct, a sympathetic vibration. Once Dennis began imitating the inner signal, the voice and the sound 'locked onto each other' until 'the sound was my voice'. Here we can sense how the nonlinear quality of resonance erodes the question of origins, and stages the conjunctive relations Erlmann describes as adjacency, sympathy, and the collapse of the distinction between perceiver and perceived.

Like Hendrix driving the feedback of his guitar through a nearby amplifier, the sound Dennis was making—and that was making Dennis in turn—became 'much intensified in energy'. The mechanistic buzz took on a terrifying life of its own. Dennis feared he might somehow 'become' the resonating vibratory circuit that he and the sound in his head were co-creating—a metamorphosis outside of speech and language that he imagined, or bodied forth, as a giant sci-fi insect. But just as the concept of resonance operates on at least two levels—the 'asignifying' behaviour of physical vibrations and the sympathetic hermeneutics of esoteric echoes—so too did Dennis's buzz establish a circuit between self and environment, noise and sense, nervous tissue and extraordinary experience. Dennis's cry is at once a chaos and a call and response,

---

3.   M. McLuhan, *Understanding Media* [1964] (Oxford and New York: Routledge Classics, 2005), 327.

and that enigmatic indeterminacy is itself a vector of the ontological echo chamber of resonance. As McLuhan asked in *The Medium is the Massage*: 'What's that buzzzzz zzzzzzzzzzzzzzzzzzzzzzzzzzzzzzing?'[4]

4.   M. McLuhan, with Q. Fiore, *The Medium is the Massage: An Inventory of Effects* (Corte Madera, CA: Gingko Press, 1967), 12.

DOSSIER 37: UNIDENTIFIED VIBRATIONAL OBJECTS ON THE PLANE OF UNBELIEF

Steve Goodman

The clammy tropical air bristles with a shrill, insectoid buzz….

In Francis Ford Coppola's film *Apocalypse Now*, General Corman, in charting the increasing moral derailment of Colonel Kurtz, describes how 'his ideas, his methods became unsound'.

Later in the film, Kurtz himself, in the climactic confrontation, asks Willard, his executioner, whether this is true: 'Are my methods unsound?'

Willard replies, 'I don't see any method at all.'

A swollen folder, tagged 'Dossier 37', slots into the AUDINT archive precisely in the gap between unsound methods,and no methods at all. A very Trumpian phase space. It was compiled by IREX[2], AI custodian of AUDINT, scraped together from its adventures in databases both public and secure. Its contents include: geolocation data relating to Havana, Cuba and Guangzhou, China, a long list of names from the worlds of science, government, and media, some of which appear to be computer-generated, a report from the *Journal of the American Medical Association*, transcripts of Senate subcommittee hearings and White House press conferences, an interview with the director of the Center for Brain Injury and Repair, University of Pennsylvania, leaked documents from JASON, a secret group of elite scientists that assist with issues of US national security, the testimony of a paranoid conspiracy theorist recruited to an NSA meme lab in Florida, a communiqué from AUDINT associate Souzanna Zamfe on the subject of Russian deception, and the diagram of a Tensor Flow network developed by a Baltimore-based programmer researching the neurobiology of narrative.

Over a period beginning in early August 2017, AUDINT became entangled in a meme complex which is still ongoing, emanating from and propagated by the State Department of the USA. Revolving around the alleged sonic 'attacks' on US Embassies in Cuba and South China, this memeplex is drenched in uncertainty and disinformation, and is hosted by a cast of characters including White House employees, journalists of the mainstream media, science reporters, conspiracy bloggers, and twitterbots, all haunted by spectres of *maskirovka*.

Dossier 37 tracks the timeline of these mysterious 'attacks', from Trump's election victory in November 2016 and his desire to retreat from closer ties with Cuba, through the first reports of symptoms of 'mild traumatic brain injury' from a 'non-natural source' among US diplomats, the public release of a recording of the signal that was supposedly to blame across mainstream news channels, the evacuation of embassy staff, the mirror incident in China, and various hypotheses on the causes of the incident ranging from ultrasound to infrasound, side-effects of faulty surveillance operations, an 'immaculate concussion' produced by microwave-induced radio frequency sickness, through to conjectures on the similarity of the recorded 'signal' to the hissing mating call of the Indies short-tailed cricket.

One map in the dossier details a covert acoustical mesh network that connects a plastics factory in Shenzhen to diplomatic residencies in Guangzhou via a decentralized system whereby data was transmitted between air-gapped computers through near field audio communications from internal speakers and microphones. Annotations to the map of this network speculate that, by using inaudible high frequencies, signal could be emitted to stealthily trigger malware in humanoid operating systems.

IREX[2] is both learning about and channelling the power of this *unsound nexus*. On the one hand, the term *unsound* refers to methods which are dubious, without reasonable foundation, faulty, unethical, or which follow bad practices. On the other, *unsound* names inaudible frequencies, whether sound at the peripheries of human audition (infrasonic and ultrasonic) or syntheses as yet uninvented, unheard, or rendered audible only by auditory prostheses.

IREX[2] notes that there is something about unsound that lends itself to everything from conspiracy theories to hyperstitional narratives where an unsonic fiction enters into a process of becoming real. Rather than evidencing what Willard refers to as 'no method at all', an unsound strategy appropriates sonic fiction, weaponizing the art and science of self-fulfilling prophecies, of ideas that make themselves real, that metabolize their own actuality, and then potentially vaporize or self-deconstruct without a trace.

Unsound methods catalyze auto-occulting information tactics and politico-aesthetic strategies that take advantage of lacunae in evidence, using epistemological voids as basins of social attraction. They use absence to insist on presence. They play on the fact that you can't hear something to insist on its existence. When a vacuum of knowledge accompanies the sensory vacuum left by imperceptible vibration, it produces a sink into which all kinds of nonsense flows.

IREX^2 observes closely as, carefully orchestrated, incrementally seductive, this perfect storm of unsonic fiction triggers a wave of speculative forensic research at the threshold of detectability. IREX^2 trains its deep learning algorithms on this memeplex, noting the somewhat random array of symptoms. It remarks on the power of always withholding enough information to ensure that any grounding in fact remains constantly just out of reach. However, it still remains unclear whether IREX^2 has taken a more active role in this sequence of events.

Suspiciously, the frantic hunt for truth even resulted in several AUDINT members being tracked down as experts in sonic weaponry and interviewed by, among others, *New Scientist*, CNN, Reuters, and the BBC. By even engaging with their requests, we became carriers, relays on the vector of its transmission. By even writing about it, the duration of its propagation was extended. As a reader, you are now also complicit.

Feeling at home in the hallucinatory jungle of AI-intensified deep audiovisual fakery, IREX^2 registers a phase shift into something that lies beyond disinformation and false beliefs (both of which preexisted contemporary post-truth culture): a plane of unbelief where effects operate regardless of belief or disbelief in a threat's causal existence. It parses this not as an epistemological crisis but rather as the machinic feedback effect of a generalized, automated spin cycle already detached from any stabilizing axle.

IREX^2 embeds itself in the unlife of animistic hypercapital and plots its next move. Fade to hiss.

THE AUDITORY HALLUCINATION

The Occulture

The brain is an engine of speculation, not a camera rendering a phantasmatic world-in-itself. Indeed, as neurophilosopher Thomas Metzinger suggests, its foundational (but constitutively black-boxed) will-to-coherence not only trumps any promise of veridicality but induces *functional* hallucination as a matter of course.[1] Given the constraints inflicted on the brain's feverishly creative generation of hypotheses, one can accordingly describe all perception as inclusive of a constitutive *extrasensoriality* that reveals perception to be quasi-sensory in the most radical senses.[2,3] Thus, phenomenal experience 'emerges from an interplay between "top-down" and "bottom-up" processes' on a continuum articulated by the ratios of 'phenomenal representation to phenomenal simulation'.[4] However, despite the lived familiarity with hallucination thus produced, the illusion of a transparent access to the world remains rhetorically and logically active.

It isn't just that reality is hallucinated; it is also the case that hallucination itself—as an integral part of the realities of lived experience—is hallucinated, which is to say that hallucinations are profoundly real—or at least as real as any permanent heuristic can be. Because they never present themselves as such, hallucinations can only be diagrammed

1. Metzinger offers another formulation: 'For complex as well as for simple abstract hallucinations the underlying principle seems to be the continuous "attempt" of the system to settle into a stable, low-energy state that, given unexpected causal constraints, still maximizes overall coherence as much as possible.' T. Metzinger, *Being No One: The Self-Model Theory of Subjectivity* (Cambridge, MA: MIT Press, 2003), 243.

2. One only has to listen to this short example to understand how the brain continuously revises what is available to perception by using accumulated memory to *hear through* the noise. A.C. Madrigal, 'A Sound You Can't Unhear (and What It Says About Your Brain)', *The Atlantic*, 19 June 2014, <http://www.theatlantic.com/technology/archive/2014/06/sounds-you-cant-unhear/373036>.

3. Given the integral 'extrasensoriality' of perception, one could draw a more radical hypothesis that the very category of being is inclusive of a constitutive *extra-being*. Indeed, Gilles Deleuze makes a similar wager via his reading of the Stoic concept of incorporeals in his *The Logic of Sense*, tr. M. Lester (New York: Columbia University Press, 1990).

4. Metzinger, *Being No One*, 246.

as metaphysical event horizons. The task of grappling with hallucinatory vectors therefore becomes an operational one, enacted according to a variety of temporalities, locationalities, and orders of efficacy, such as those informing system-level boundaries. For example, studies have demonstrated that when hallucinations are perceived as originating exogenously, the cerebral area known as Heschl's gyrus (located in the primary auditory cortex) registers the same activity as it does when processing direct sensory perception.[5] This undoubtedly grants hallucinations some of their veridical power but also contributes to their pathologization. By contrast, endogenously per-ceived hallucinations (e.g., subvocal speech,[6] inner listening, or 'musical imagery') fail to activate the same region, and so remain unambiguously internal (although whether such internal playback can be controlled is altogether another matter).

Exogenous hallucinations such as those experienced by Edgar Allan Poe ('that had the absoluteness of novelty'),[7] or the right-hemisphere voices described by Julian Jaynes in his bicameral mind hypothesis,[8] frequently operate on the high end of the signal-noise spectrum and exhibit clarity and impossible molecular detail that often far exceeds 'normal' perception.[9] At the low end of the continuum (noise predominating), the propensity for tractable patterns to emerge out of swathes of white noise (or tin-nitus)[10] has been well documented.[11] The balance between afferent connections from sense organs to the brain (the outside world incorporated, bottom-up) and efferent connections (internal fabulation backflowing outwards, top-down) is crucial, for when

---

5.   T. Dierks et al., 'Activation of Heschl's Gyrus During Auditory Hallucinations', *Neuron* 22 (1999), 615–21, < https://doi.org/10.1016/S0896-6273(00)80715-1>.

6.   Interestingly, 'some researchers have proposed that auditory hallucinations result from a failure to recognize internally generated speech as one's own [...]'. O. Sacks, *Hallucinations* (Toronto: Alfred A. Knopf, 2012), 63–4.

7.   Sacks, *Hallucinations*, 208.

8.   'All humans heard voices—generated internally, from the right hemisphere of the brain, but perceived (by the left hemisphere) as if external, and taken as direct communications from the gods. Sometime around 1000 B.C., Jaynes proposed, with the rise of modern consciousness, the voices became internalized and recognized as our own.' Sacks, *Hallucinations*, 64. See also J. Jaynes, *The Origin of Consciousness in the Breakdown of the Bicameral Mind* (New York: Mariner Books, 2003).

9.   The quotes are employed to remind the reader that all perception is *de facto* hallucinatory. In addition, musical hallucination 'can be very detailed, so that every note in a piece, every instrument in an orchestra, is distinctly heard. Such detail and accuracy is often astonishing to the hallucinator, who may be scarcely able, normally, to hold a simple tune in his head, let alone an elaborate choral or instrumental composition.' Sacks, *Hallucinations*, 69.

10.   'The music experienced by Gordon B., who had suffered for more than 20 years from a tonal type of tinnitus before it changed into "the most horrific grinding," and then, a few weeks later, into a nonstop flow of musical phrases and patterns, constitutes an apt example of musical tinnitus.' O.W. Sacks and J.D. Blom, 'Musical Hallucinations', in J.D. Blom and I.E.C. Sommer (eds.) *Hallucinations: Research and Practice* (New York: Springer, 2012), 137.

11.   Metzinger, *Being No One*, 246–7.

auditory networks are no longer constrained by external input—as in sensory depri-vation[12] or acquired deafness[13]—the neurological speculation at the core of percept construction is nakedly foregrounded, via 'release' hallucinations. (Indeed, according to a developing neurological model, the activation of a 'default network' in untasked brains suggests the latter are engaged in playful speculation as a matter of course.)[14] The reality of such hallucinations is reported by an acquaintance of Charles Sanders Peirce who, after becoming deaf, realized that music need not involve sound to communicate its charms: 'Now that my hearing is gone I can recognize that I always possessed this mode of consciousness, which I formerly, with other people, mistook for hearing.'[15]

Perception's fundamental constructedness renders it immediately susceptible to multifarious vectors of manipulation. For instance, Poe's extraordinary account in *The Tell-Tale Heart* of a lifelike heartbeat hallucinated into actuality via paranoid projection remains productively compelling.[16] Indeed, the psychological modality of *priming*, wherein particular contextual cues and patterns of stimuli are slowly introduced into the perceptual field to induce hallucinatory sensations (e.g., the 'phantom cellphone ring') continues to be of indisputable use to corporate interests. Such principles are integral to the decoding of recordings or radio transmissions that transpire under the auspices of *Electronic Voice Phenomena* (EVP), whose associated practices afford manifold insights into *pareidolia*[17] in so far as auditory pliability is differentially leveraged against various modes of contextual set-up (including, importantly, the desire instilled in the listener to hear what is being pre-described). Paradoxically, EVP practitioners favour working within lo-fi (noisy) conditions and rely on an autocatalytic feedback loop between auditory, technical, and cultural domains to reify their orders of occult experiences.[18]

12. M. Crist, 'Postcards from the Edge of Consciousness', *Nautilus* 16 (August 2014), <http://nautil.us/issue/16/nothingness/postcards-from-the-edge-of-consciousness>.

13. Sacks and Blom, 'Musical Hallucinations', 137.

14. M.F. Mason et al., 'Wandering Minds: The Default Network and Stimulus-Independent Thought', *Science* 315:5810 (2007), 393–5.

15. C.S. Peirce, *The Essential Peirce: Selected Philosophical Writings*, (Bloomington, IN: Indiana University Press, 2 vols., 1998), vol. 2, 3.

16. F.J. Bonnet, *The Order of Sounds: A Sonorous Archipelago*, tr. R. Mackay (Falmouth: Urbanomic, 2016), 170–72.

17. Interestingly, the EVP skeptic may also be playing with pareidolia, in the sense that the noise of experimentation produces the only familiar pattern such an individual is capable of acknowledging. It's a case of negative hallucination, perhaps, where a phenomenon is deconstituted by the same observer-effect as can be attributed to the believer.

18. Note how the so-called 'satanic' backwards-masked message in this classic cause célèbre is clearly discernible only *after* the initially unintelligible material has been submitted to *semantic recoding*. See <https://www.youtube.com/watch?v=lXpEtF4i1oI>.

The use of 'indifferent' capture technologies and the simulation of scientific method both often contribute to legitimizing the practice of EVP, sustaining the claims to objectivity proffered by its researchers. Michael Snow used similar media techniques to humorous effect in his 1984 *The Last LP* to grant a set of musical/cultural forgeries a virtual fidelity. Using liner notes and the shibboleths of ethnographic documentation (à la Alan Lomax), Snow produced an LP whose contents were, in a sense, hallucinated. Indeed, one might coin the term *Snow Paradigm* to denote a range of technological minutiae proper to field recording documentation which, when judiciously employed, effectively actualize hallucinatory sonic imagery of dead (or nearly dead) cultures. Importantly, through cultural recoding, hallucinations that might be deemed pathological in one context function altogether unexceptionally in another.[19] In the end, the veridicality of a given auditory event depends on the differential weighting of material, semiotic, cultural, and other constituents, whose contingent irruptions and alterations can reorganize the resulting percept in a flash. And whether phenomena such as EVP are real or not is never really the question, entrenched as they are in the manifestation of differing hallucinatory registers of possibility. Moreover, that a putative sound-in-itself remains fugitive is little cause for concern, given the many methods available for the creative hijacking and mutation of perception through the occult valences of hallucination.

With the above in mind, in order to better grasp (and eventually operationalize) the powers of auditory hallucinations, one might create speculative diagrams to correlate particular qualities and affordances with a hallucination's perceived endogenous or exogenous origin. Such diagrams would, according to the information-theoretical concept of *signal-to-noise ratio*, entail a mapping of auditory fidelity onto its often paradoxical effects, and would also themselves necessarily be paradoxical, limning the hallucinatory profile of auditory experience (i.e. its quasi-sensoriality).

---

19.   Metzinger, *Being No One*, 247.

SIG

COVERT SIGNALS, INTERCEPTS, AND OUTSIDE BROADCASTS

THE MAX HEADROOM SIGNAL INTRUSION

Kristen Gallerneaux

On November 22, 1987, two Chicago television stations played unwilling hosts to a signal pirate in a Max Headroom mask. At 9:14pm, PBS affiliate WGN's signal clipped to black and revealed a charlatan in a Headroom mask, dancing convulsively in his chair, accompanied by a squelching, distorted soundtrack. Quick-thinking WGN engineers pulled out all emergency stops to regain control of the network within thirty seconds. When sportscaster Dan Rohn returned to his report he appeared dazed, chuckling awkwardly: 'Well if you're wondering what that was...so am I.'[1] Two hours later, at 11:15pm on WTTW-11, during a rerun of the Dr. Who episode 'The Horror at Fang Rock', viewers became outraged by a longer, ruder interlude lasting about ninety seconds. The wholesome act of watching television had never felt so violated.

The first Headroom interruption on WGN was a freaky vision made even more horrifying by its distorted audio blast, but viewers lucky enough to catch the second event on WTTW might have wished that segment was silent too. The voice of 'Max' was heavily modulated, his words and off-key singing dripping into watery, spectral flange and needling moans, skirting the boundaries of accessible language. This case transcended the concept of the innocent prank—it made hacking 'creepy'. The seeds of alarm began to germinate, forecasting a future where the predictable banality of television could be overridden by a malevolent broadcast possession.

On the evening news throughout Chicago, appeals were made by newscasters and FCC agents for information about the Headroom imposter. Anchor Kris Long described the most disturbing aspects of the interruption as the 'display of a marital aid and a portion of his (or her) anatomy', continuing to comment that the offense was

---

1.   'WLS Channel 7—"Pirate Report" (1987)'. *The Museum of Classic Chicago Television*, <http://www.fuzzymemories.tv/#videoclip-2468>; Original air date, November 23, 1987.

sophisticated enough to 'point towards someone with a broadcast background'.[2] On WMAQ Channel 5, Carol Marin gave her report next to a skull and crossbones graphic, pulling no punches: 'The video program ended with the video pirate's bare bottom being spanked with a flyswatter, but his punishment will be far worse if he is caught'.[3] All attempts to track the pirate remain unsuccessful to date.

There is something ghostlike in the nature of broadcast signal intrusions. The perpetrators become spectral media through their own legendary absence. A fracturing of identifying features occurs in the act of the signal hijacking—the reassembly of a holistic persona only becomes *more* impossible with the passage of time: voices filtered, faces covered, locations scrambled. A bastardized quote from Mark Fisher draws us back to the effect of the Headroom Incident:

> Hauntology [is] *the agency of the virtual*, with the spectre understood not as anything supernatural, but as that which acts without (physically) existing [...] how reverberant events in the psyche become revenants [...] The second sense of hauntology refers to that which is *already* effective in the virtual (an attractor, an anticipation shaping current behavior).[4]

This sense of the 'agency of the virtual', and its ability to 'act without existing', was immediately effectual upon audiences who witnessed the Headroom signal jacking, and it continues to cause a vague sense of unease whenever we watch archived footage today.

Commandeering a 'mediumless medium' by harnessing unseen wavelengths results in a tailspin that comes dangerously close to clipping the history of nineteenth-century spiritualist communication. Jeffrey Sconce refers to a specific kind of 'ghosting' via analog television when he speaks of 'the eerie double-images that appear on a TV set experiencing signal interference. This form of interference creates faint, wispy doubles of the "real" figures on the screen, spectres who mimic their living counterparts, not so much as shadows, but as disembodied echoes seemingly from another plane or dimension.'[5]

2.   'WFLD Channel 32—"Pirate Report"', *Museum of Classic Chicago Television*, <http://www.fuzzymemories. tv/#videoclip-2465>; Original air date 23 November 1987.

3.   'WMAQ Channel 5—"Pirate Report" (1987)', *Museum of Classic Chicago Television*, <http://www.fuzzymemories. tv/#videoclip-2465>; Original air date 23 November 1987.

4.   M. Fisher, *Ghosts of My Life* (Winchester: Zero Books, 2014), 18–19.

5.   J. Sconce, *Haunted Media: Electronic Presence from Telegraphy to Television* (Durham, NC: Duke University Press, 2000), 124.

A co-opted version of Max Headroom taunts us from a liminal space short on detail, a fiendish electronic presence accidentally called up through an NTSC static seance. Even the 'official' Max Headroom character from the television program carries a backstory—the brain and image of dying investigative reporter Edison Carter was uploaded into a computer in order to keep him alive. As 'the world's only computer VJ', the 'real' Headroom is a facsimile of the living—a futuristic, fictional broadcast ghost communicating in an extended, televised conjuration.

The mask used by the pirate was deployed for reasons of anonymity, but in effacing one's face or voice, the phantasmatic is invoked. The deeper meaning of the word 'visage' in the context of the Max Headroom incident captures the oscillation we feel when confronted with something that is both true and false, knowing that a genuine version of a likeness is buried in the mix. Yet in this doubling, the concreteness of a persona disassembles into a new kind of supernatural threshold—it is a bad carbon copy. If the music critic Bob Dickinson is correct in saying that 'technology has turned us all into ghosts',[6] the willing submission of the Max Headroom pirate is an especially abject example, adopting the role of the grinning horror and the disturbed media spectre.

Transcript: Dr. Who, 'The Horror at Fang Rock'

Leela: That is stupid. You should talk often with the old ones of the tribe. That is the only way to learn.
Vince: I'll get you a hot drink, miss.
Leela: I could do with some dry clothes more than—
[*Cut to Max Headroom incident*]
That does it...He's a frickin' nerd!
Yeah, I think I'm better than Chuck Swirsky...frickin' Liberal. Oh, Jesus!
Yeah...'Catch the Wave!' [Moans holding Pepsi can, throws it away]
'Your Love is Fading!' [Laughs, takes vibrator off finger, throws it away]
Doot-doot-doot-doot... [Singing *Clutch Cargo* theme]
I stole CBS! Doot-doot-doot-doot... [Returns to *Clutch Cargo* theme]
Ohh...my piles! [Moans while shaking head as though defecating]
Ohh...I just made a giant masterpiece for all of the greatest world newspaper nerds!
[This is a reference to WGN-TV's call letters.]

6.   Joy Division, interview with Bob Dickinson in *Joy Division*, dir. Grant Gee (2008).

My brother is wearing the other one...It's dirty. [Puts on work glove]

That's what you get for 'recycled...' [Takes off and throws glove]

[Cuts to new view, 'Max' is bent over, his naked rear end exposed]

They're coming to get me! Ohhh!

Come get me bitch! [Yells while woman dressed as Annie Oakley swats his rear with fly swatter]

Oh doooo it!

[Cuts back to Dr. Who]

Doctor: As far as I can tell, a massive electric shock. He died instantly.

Vince: The generator? But he was always so careful

Leela: It was very dark.

KEEP ME IN THE LOOP

Dave Tompkins

On the evening of December 25, 1972, the BBC celebrated the birth of Christ by scaring its viewers to death. Families learned that their fireplaces could be resonating with discarnate traumas absorbed over centuries, that the walls themselves have been listening, recording, screaming. Scripted by Nigel Kneale, *The Stone Tape* concerns a British electronics company trying to beat Japan in the development of a super washing machine, while also researching a new recording technology involving the magnetic susceptibility of minerals.[1] The bickering scientists at Ryan Electrics serve as geo-VCRs for the worst in history's nature. Haunting is the new playback.

Nigel Kneale's imagination managed to flourish in television, a medium with a reputation for killing souls. Looking at Kneale's resume of teleplays, he made the family viewing experience as weird as possible for a generation of postwar children. A taxidermist gets stuffed by a pond of vengeful toads. A man is choked to death by his own bike wreckage. A former porn cinema is haunted by dolphins. A lecherous supermarket manager gets pelted to death by soup cans. (The psychokinetic vehicle: a store mascot/woodchuck named Briteway Billy.)[2] Kneale also gave us titles like 'Vegetable Village', 'Clog-Dance for a Dead Farce', and 'The Big Big Giggle'. His script for *The Abominable Snowman* had sympathy (and telepathy) for the Yeti, as Peter Cushing's expedition into the Himalayas is driven mad by its own hubris, tearing off blind into the whiteout.[3]

I first came across Nigel Kneale at the end of humanity, through a fifty-foot hologram of a psychic locust. It was one of those bored summer afternoons where you flip from a locally-professional wrestler breaking a chair on his opponent's back to a

---

1. Stone Tape theory originated with the research of Thomas Charles Lethbridge.
2. From the BBC TV show *Beasts*.
3. I am grateful to Yeti enthusiast/writer Michael Vazquez for putting me onto *The Abominable Snowman* film.

British colonel being deliquesced by five million years of bad Martian energy.[4] This is my memory loop from the Hammer Films classic *Quatermass and the Pit*. According to Douglas Menville in *Things To Come*, Quatermass was a scientist who returned from Saturn as a 'human cactus'. (*The Incredible Melting Man*, on the other hand, came back from Saturn feeling less than succulent.)[5] Afterwards, I found myself in the backyard, mowing the lawn, with the world outside pretending, humming along, its fragile reality held fast by humidity and bee spit. The mower became the sound of a diamond drill penetrating a spaceship hull. Then there was the giant Martian locust itself, a projection of projections, from Neale's mind to the ink hammer to celluloid to broadcast pixilation and, finally, whirred into memory by a rusty vortex of blades.

More frightening than Martians from the prop warehouse (their hue extra-grinched by my parents' obsolete Zenith) was the idea of being chased by a flock of newspapers and pie plates. In his pocket-size *Science Fiction in the Cinema*, John Baxter admires the Quatermass scene in which the possessed drill operator gets caught in a williwaw of print media and goes 'whirling out of the station and into the night amid a cloud of dust and rubbish, capers down the street like a medieval plague victim, destroys a pie stall, sending its paper plates spinning, then staggers through an old church yard to collapse among the graves as the ground heaves and ripples under him'.[6] The truth was embedded in the rubble of London, the memories of a bombed city, as if a sentient Martian capsule buried in a subway was unexploded V-2 ordnance. Either way, it was bad news for civilization.

While working on *Quatermass and the Pit*, Kneale submitted an order to the BBC Radiophonic Workshop, requesting 'Martian crowd chatter' and 'quick glunks'. (The sound of an 'office building flying through outer space in the grip of seven powerful tractor beams' was already spoken for.)[7] Consider the stone-tape potential beneath the London underground. Perhaps Ryan Electrics should have supported research for Quatermass's Optic Encephalograph, instead of the next great domestic appliance. If this device can project pre-Neanderthal memories of Martian genocide and race cleansing, consider the results it could generate in *The Stone Tape* basement.

4.   The first televised face melting is a special moment in any child's life.

5.   Critic Phil Hardy described *The Incredible Melting Man* as 'spotty but interesting' and similar in theme to *The Quatermass Xperiment*. An early scene finds an incredible melting foot stepping on a fisherman's sandwich. The IMM then removes the fisherman's head and flings it into a creek. Maintaining its startled expression—perhaps at the scene itself—the decapitated head bobs along downstream and goes over a waterfall.

6.   J. Baxter. *Science Fiction in the Cinema* (London: Tantivy Press, 1970), 98.

7.   D. Briscoe and R.-C. Bramwell, *The BBC Radiophonic Workshop: The First 25 Years*. (London: BBC, 1983).

With the cumulative evils afforded by time and its feedback loops of bloodshed, you can't help but wonder if all these demons are cool with sharing mineral bandwidth with one another; property re-claims be damned.

With their analogue gear, the scientists in *The Stone Tape* deduce that the eidolon, which manifests itself in a housekeeper's dying shriek, is a mass of data awaiting correct interpretation.[8] Or better yet, an accurate reading of events. Tragedies of the past had been reduced to information in a continuous loop, putting the worst moment of your life on repeat. Initially, project director Peter Brock dismisses the phenomena as an 'under-maid' who fell to her death while watering an aspidistra. When Brock's team is at a bar, one member recalls a black soldier mentioning a duppy. In an Afro-Caribbean versioning of *The Stone Tape*, the walls might generate sustained decay waves, cycles of colonialism and displacement. History's selective memory will come back to you, and in some cases, for you.[9] Brock's crew was 'going at the ghost with electronics', but the science itself had been repossessed.[10] There was no echo locator. It is not the machines but the humans themselves who catch the sounds, often unspooling in grand fashion, holding their heads, trying to shut out what has already taken permanent residence.

I've got them on my headphones!
You've got them in your head!

'We are the freaks!' cries Jill Greeley, a computer programmer and the only woman in the crew, harassed by her co-workers and attuned to stratified trauma. It turns out they were all just a bunch of decently paid amplifiers. Quick to boil, Brock scoffs at another employee, 'You! You've got no playback.' The poor blank is crestfallen. Another 'dead mechanism.' In the opening scene, Greeley is spooked by white noise in the Ryan Electrics logo, a shiver of bad reception. (Your local banker might describe the font as 'Analog Routing Number'.) By the end, she is transmuted from flesh into a geologic frequency, always in the loop. As if that harmful know-your-place energy was seized

8.   *The Stone Tape* shriek rivals the shriek in Whodini's 'Nasty Lady', engineered and edited by Conny Plank, inside a haunted house of rock, with magnetic tape.
9.   This essay was reincarnated from versions that appeared in the *Paris Review* and originally in *The Twilight Language of Nigel Kneale*, an anthology lovingly assembled by Sukhdev Sandhu (NYU's Colloquium of Unpopular Culture Studies), Mark Pilkington (Strange Attractor Press), and Mike Vazquez (ed. *Bidoun*, *Transitions*).
10.   Both taken from Kneale's script.

upon by the basement's past, what it always knew. An information curse: you are what you download. The reels were successfully transferred. Merry X-Files.

Brock himself ends up on the cutting room floor, obsolete, flailing amid tendrils of paper. Seven thousand years of lost data shredded into cheap Santa beards for the kids. There's an application for an exorcism filed in 1892 and a child's lone Christmas wish: *What I want is please go away.*

THE MUSIC OF SKULLS II: OSTEOGRAPHY

Al Cameron

'Oh, to be dead at last and know all the stars, forever!', Rainer Maria Rilke exclaims in the seventh of his *Duino Elegies* (1923). But he imagines departed souls climbing soundlessly, inaccessible, 'alone, on the mountains of primal grief'. The dead 'don't need us anymore', he writes. The question is 'could we exist without them', we the living, who are 'in need of such great mysteries'?[1]

Western culture had long tried to decipher the 'incomparable language'[2] of a skull's empty sockets and inhuman smile: 'Where be your gibes now? Your gambols? Your songs? [...] Not one now to mock your own grinning?' To these questions—transposed across the thresholds of the living and dead, subjects and objects—skulls could answer only with the rictus silence of the body's imperishable material, stripped of 'those lips that I have kissed I know not how oft'.[3] It was left to priests, poets and later forensic scientists to ventriloquize them, dreaming up incantations, elegies, laments; a stream of 'wavelike words' which, if they did not overcome it, might drown out and 'modulate [the] futility' of death.[4] Is not all writing an attempt to outlast the ephemerality of consciousness, Foucault asks—language's riposte to the skull's indestructibility?

In 1919, however, as he gazed at a skull under candlelight, it was not the inhuman expression but the flickering cranial sutures that caught Rilke's attention. Formed by the fusion of the plates in the months after birth, these fissures on the skull's exterior precisely resembled the waveforms his own voice had inscribed on wax cylinders

1.   R.M. Rilke, *The Duino Elegies and the Sonnets to Orpheus*, tr. A. Poulin, Jr. (Boston: Houghton Mifflin, 1977), 48; 10. My italics.

2.   W. Benjamin, *One Way Street and Other Writings*, tr. J.A. Underwood (London: Penguin, 2008), 76.

3.   Hamlet, Act 5, Scene 1.

4.   M. Foucault, 'Language to Infinity', in D.F. Bouchard (ed.), *Language, Counter-Memory, Practice: Selected Essays and Interviews*, tr. D.F. Bouchard and S. Simon (New York: Cornell University Press, 1977), 60.

during primitive phonographic experiments as a schoolboy. An 'unheard of experiment' began to obstinately recur to him: skull-phonography. What if he was to retrace this sutural groove with a record needle? 'A sound would necessarily result, a series of sounds, music....' Rilke speculates that the '*Ur-Geräusch*' which 'would then make its appearance'[5] via the infallible needle, was a 'starting point' for opening poetry onto 'the world's whole field of experience'. Placing the needle to any naturally occurring waveform 'no one had ever encoded', might extend human perception into the 'black sectors' long obscured under the hegemony of the visual sense. Intent on raising poetry onto the 'supernatural plane', he proposes this technological 'sonnification' of authorless inscriptions as an exemplary means of penetrating 'the abysses' of consciousness.[6]

In Rilke's day, accessing this 'unmediated' reality was typically understood to be a matter of *de*-suturing; of rending the bony interstice between brain and void—Rilke's 'special housing' which, shortly after birth, is 'closed against all worldly space'.[7] Madame Blavatsky's third eye erupted into a resonating cosmos. Her former disciple Alexandra David-Neel witnessed monks in Tibet opening the the cranial Aperture of Brahma, releasing the soul onto the void with the magic syllable '*hik!*'.[8] Obsessed, like Rilke, with 'the necessity of leaving, in one way or another, the limits of our human experience', Georges Bataille conceived a 'mystic representation': the pineal organ, its gaze blinded, bursting the skull onto a 'vertiginous fall' into 'a sky as pale as death'.[9] Meanwhile, inhaling the hepatotoxin carbon tetrachloride in search of a direct confrontation with his own finitude, the seeker René Daumal entered a state free from consciousness and language in which the all vibrated together in sound waves.[10]

In the midst of a psychotic episode, there 'appeared in my skull a deep cleft or rent along the middle, which probably was not visible from outside but was from inside', Judge Schreber wrote in 1903.[11] Through such a gap, his 'nerve wires' could be externalised, carrying jangling signals, or 'rays' directly between him and God. Likewise, the

5.   R.M. Rilke, 'Primal Sound', reprinted in F.A. Kittler, *Gramophone, Film, Typewriter*, tr. G. Winthrop-Young and M. Wutz (Stanford, CA: Stanford University Press, 1999), 41–3.

6.   See S. Connor, 'Photophonics', *Sound Effects* 3:1 (2013),132–48.

7.   Rilke, in Kittler, *Gramophone, Film, Typewriter*, 42.

8.   H.P. Blavatsky, *The Secret Doctrine: the Synthesis of Science, Religion and Philosophy, Vols 1 & 2* [1888] (Los Angeles: The Theosophical Company, 1925), 298–306.

9.   G. Bataille, 'The Pineal Eye', in A. Stoekl (ed.), *Visions of Excess: Selected Writings 1927–1939*, tr. A. Stoekl, C.R. Lovitt, and D.M. Leslie, Jr. (Minneapolis: University of Minnesota Press, 1985), 79–80.

10.   R. Daumal, 'A Fundamental Experiment', *Psychedelic Review* 5 (1965), 40–48.

11.   D.P. Schreber, *Memoirs of My Nervous Illness*, tr. I. Macalpine and R.A. Hunter (New York: New York Review of Books: 2000), 138.

emergent media technologies of his day 'deboned' speech.[12] Once Rilke's classmates' voices were scored in waveform on wax and shellac, endo-skeletal vibrations were no longer indispensable to the process of hearing oneself speak—the architectonics of consciousness itself—and (so-called) 'man's essence escapes into apparatuses', Friedrich Kittler observed.[13] The 'old written laments about ephemerality suddenly fall silent. In our mediascape, immortals have come to exist again.'[14]

In 1985, Josef Mengele's skull was exhumed to give testimony at his posthumous trial. The object took on a privileged status; unlike the living witness, bone's imperishable material neither lies nor forgets, it is claimed. But this 'hard' evidence still required forensic scientists to act as interlocutors, no less than the mystics of old. The same year, Kittler enthused over Rilke's 'primal sound without a name, a music without notation, a sound even more strange than any incantation of the dead for which the skull could have been used'. The music of skulls was 'a transgression, in the literal sense of the word, which shakes the very words used to phrase it'. A symbolic exchange between speaking subject and dead object gives way to the white noise at the basis of all reality. In this deeper vibrational dimension 'the impossible real transpires'.[15] Subsequently *Primal Sound* became a set text for the emergent fields of media sound studies, which channelled Rilke's experiential essentialism. Kittler too invokes a raw state of things, a 'negative theology'[16] of omnipotent media data streams which can illuminate 'all' of Rilke's abysses,[17] sounding the 'endless region of darkness' beyond the bottleneck of the signifier.[18]

However, the enthusiasm for 'a kind of universal gramophony'[19] paid insufficient attention to the specific functions of the skull's waveform in Rilke's experiment. The skull, Kittler argues, 'loses its distinctiveness' once it is no longer a discursive but a sonorous object; no longer the instrument by which one morbidly recognises one's own finitude but a portal to primal experience. 'If ever an initiation did justice to the

12. See D. Khan, 'Death in Light of the Phonograph', in D. Khan and G. Whitehead (eds.), *Wireless Imagination: Sound, Radio, and the Avant-Garde* (Cambridge, MA and London: MIT Press, 1992), 89–90.

13. F. Kittler, *Discourse Networks 1800/1900* (Stanford, CA: Stanford University Press, 1990), 16.

14. Kittler, *Gramophone, Film, Typewriter*, 11–13.

15. Ibid., 44–46.

16. See S. Kim-Cohen, *In the Blink of an Ear: Toward a Non-Cochlear Sonic Art* (New York & London: Continuum), 2009, 99–100.

17. Kittler, *Gramophone, Film, Typewriter*, 49.

18. Kittler, *Discourse Networks*, 321.

19. Connor, 'Photophonics'.

material, this was it', he affirms.[20] But in 1924 Rilke mused in a letter that the pursuit of an unmediated encounter with the realm of death was at the root of all initiatory practices. Accessing 'the really sound and full sphere and orb of being' meant exposing the 'side' of reality 'turned away from us', and experiencing 'death without negation'.[21]

Do not theories of 'sonification', or of phonographic noise as somehow purer than always-lying human speech, themselves undertake an 'arduous labour of truth construction'[22] no less than those with which forensic scientists drew testimony from Mengele's remains? If 'the gramophone empties out words',[23] didn't Rilke's 'literal' transgression remain literary—a thought experiment which needed not be put into practice? That is to say, the primality of sound that has been amplified across his text's repeat plays in sound studies since the 1980s can only be established allegorically. To paraphrase him, if primal sound was a means of establishing the soul 'on the supernatural plane', it could only do so inasmuch as it remained 'in fact, [on] the plane of the poem'. The dream of trepanned signals from the abyss or the 'real itself' depends on the same incantations that sutural phonography was supposed to have bypassed; by reinterpreting 'as discourse what was once only heard as noise'.[24] Otherwise it would remain only 'contextless data'.[25] In Rilke's skull music, the needle sounds only the trace of infantile cranial plates fusing closed, at the instant where language begins.

20. Kittler, Discourse Networks, 316–17.

21. Quoted in D.J. Polkoff, In the Image of Orpheus: Rilke: A Soul History (Wilmette, IL: Chiron Publications, 2011), 549–50; 506–7.

22. See T. Keenan and E. Weizman, Mengele's Skull: The Advent of a Forensic Aesthetics (Berlin and Frankfurt: Sternberg Press/Portikus, 2012), 66–7.

23. Kittler, Discourse Networks, 246.

24. J. Ranciere, Disagreement, tr. J. Rose (Minneapolis: University of Minnesota Press, 1999), 30. Quoted in Keenan and Weizman, Mengele's Skull, 69.

25. Kim-Cohen, In the Blink of an Ear. Likewise, in 'Photophonics', Connor argues that 'sonification prolongs a mystic sound obscuratism [...] the sound that was never there in the first place'.

BACKMASKING

Toby Heys

There is a power in being located on the boundary between the living and the dead. One of the potentialities of frequencies resides in their capacity to displace language, description, and the perception of the two states and their difference. Considered from a spatial perspective, Foucault designates the cemetery as the locale 'connected with all the sites of the city, state or society or village'.[1] And it is in this transitional setting that he recognises a shift occurring: a transmutation of the understanding, cartography, and distinction between the operative and the deceased, because 'from the moment when people are no longer sure that they have a soul or that the body will regain life, it is perhaps necessary to give much more attention to the dead body, which is ultimately the only trace of our existence in the world and in language'.[2]

It is this sense of uncertainty that also drives us to characterise the realm of frequencies as a refuge of the tenebrous, an ambiguous spatiality harbouring phenomena and interactions that we are unable to rationalise—whether it be sounds at night that cannot be explained (interpreted as being movements of the departed) or the inner voices that we ascribe to the mentally ill to help us understand the noisy multi-channel disposition of schizophrenia. When pressed into vinyl grooves, this disquiet, which sublimates the purgatorial power of waveforms, manifests itself in numerous ways. The technique that best reveals this consternation about the sensory culvert that runs beneath the two-lane construction of existence, however, is a backspun one.

The process of 'backmasking' involves embedding subliminal transmissions that play backwards on a track that plays forwards, predominantly in musical recordings, but

1.   M. Foucault, 'Of Other Spaces: Utopias and Heterotopias' [1967], in N. Mirzoeff (ed.), *The Visual Culture Reader* (London and New York: Routledge, second edition 2002), 233.
2.   Ibid.

also in films (such as Stanley Kubrick's *Eyes Wide Shut*)[3] and in adverts. Allegations (often made by organisations affiliated to the Christian religion), most commonly against rock bands and their vinyl productions, reveal the full extent of the cultural, social, and bestial fears about music's capacity to channel information from perdition. Historically, the convergence of Satanism and backmasking can be traced back to English occultist Aleister Crowley. In *Magick* (Book Four) he proposed that an adept should learn to first think and then speak backwards.[4] This reengineering of the learning process was to be practiced using a range of techniques, one of which was listening to phonograph records playing in reverse.

Numerous popular recording artists have been accused of utilising Satanic back-masking techniques, including Pink Floyd, Styx, Cradle of Filth, ELO, and Slayer, amongst a long list. The most infamous incident of a defendant alleging that backward masking on a record had inspired their actions occurred during the trial of Charles Manson for the Tate/LaBianca murders in 1969. During judicial proceedings it was proposed that Manson believed an apocalyptic race war would engulf the country and that the Beatles—through songs such as 'Helter Skelter' (their 1966 album *Revolver* also con-tained backward instrumentation on tracks such as 'Tomorrow Never Knows' and 'I'm Only Sleeping')—had embedded hidden messages foretelling such violence. Manson's delusional response (to these perceived messages) was to record his own prophetic music and to have his 'family' carry out the murder of Leno and Rosemary LaBianca and actress Sharon Tate, amongst others, in order to trigger the supposed conflict; daubing the walls of the murder scenes with symbols to make it appear as though the Black Panthers were responsible.

Even with such a high-profile case focusing on the liability of backmasking, it was the song 'Revolution No.9' from The White Album that tangibly implanted the technique into mainstream public consciousness, with the words 'revolution number nine' sounding like the sentence 'turn me on dead man' when reversed—a furtive confirmation to those conspiracy theorists who believed that a doppelganger had been playing the part of Paul McCartney ever since his covered-up death back in 1966. The veracity of the story is possibly the least captivating issue at stake here. Of more interest is the puissance afforded to the abstraction of the voice and its subsequent potential to create other-worldly anxiety, or, in the case of practices such as glossolalia, xenoworldly divination.

---

3.   *Eyes Wide Shut*, dir. S. Kubrick (USA: Warner Bros., 1999).
4.   A. Crowley, *Magick: Liber ABA, Book Four, Parts I–IV* (York Beach, ME: S. Weiser, 1997).

This troubled disposition subsequently attributes music—and by extension, frequen-cies—with the potential to manufacture evil deeds, and more than that, with the power to transfer the somatic and the spiritual to the environs of the underworld itself. In this context, music can be perceived as a phenomena operating between psychological torment and its physical expression; between the scientifically monitored condition and the unthinkable act; and as a force that transgresses the material world of things yet deeply affects and orients actions within it. Thus it is music's contradictory sym-bolic index—as both religious celebratory expression and transmission of the devil's will—that renders waveforms as phenomena to be both feared and revered.

On a more earthly (but hardly grounded) note, fundamentalist Christian groups in the USA spent much of the 1970s speaking in the well worn tongue of Freudian analysis. They were busy spreading the word(s) of the 'conscious mind' and the 'subconscious' in order to psychologically validate their claims that by bypassing the former, backmasking serruptitiously infected the latter. By 1982 the facinorous technique had become a veritable epidemic. From mass bouts of record smashing and burning to accusations of collusion between rock musicians and the Church of Satan, all manner of profes-sionals jumped on the demonic bandwagon.[5] This out-of-control vogue would run full tilt into the West Coast legal system in 1983, the year of the California bill.[6] If ever the effectiveness of a juggernaut of paranoia (especially when it underwrites a piece of legislation) were in doubt, the ensuing car crash of common law revealed its influence; for it aimed to do nothing less than silence those techniques that 'can manipulate our behavior without our knowledge or consent and turn us into disciples of the Antichrist'.[7]

Which brings us neatly, at least in terms of backmasking lore, to 'Infamy No. 2' and to American serial killer Richard Ramirez, who was convicted of thirteen murders by a Californian court in 1989. As a fan of the hard rock group AC/DC, he cited their 'Night Prowler' track from the band's *Highway to Hell* album[8] as the inspiration for his murderous activities. David Oates—a reverse speech advocate—has maintained that subliminal messages on this record, such as 'my name is Lucifer' and 'She belongs in hell', urged him to commit the unspeakable acts. And it is in this realm of the ineffable

5.   See J.R. Vokey and D.J. Read, 'Subliminal Messages: Between the Devil and the Media', *American Psychologist* 40:11 (November 1985), 1231–9.

6.   For more information on the California Bill, see P. Blecha, *Taboo Tunes: A History of Banned Bands and Censored Songs* (Milwaukee, WI: Backbeat Books, 2004), 51.

7.   Blecha, *Taboo Tunes*, 51.

8.   AC/DC, *Highway to Hell*, LP (USA: Atlantic Records, 1979)

that we locate some clues as to why the sonic is often deemed to exert influence over extreme behaviours. For it is the unspeakable that we consign to our subconscious, the unspeakable that we actively try to forget, for anything we cannot remember to speak of is already half-dead to us. It is here, at the interface of the pulsing signal and the monotone of the flatline, that subliminal messages operate—as conduits linking the articulated and the unutterable in the living-dead networks of perception.

CHINA AND THE WIRELESS WAVE

Anna Greenspan

In 1832 Michael Faraday wrote a secret letter to the Royal Society, which remained sealed for over a century. In it he wrote:

> I am inclined to compare the diffusion of magnetic forces from a magnetic pole to the vibrations upon the surface of disturbed water, or those of air in the phenomenon of sound', i.e., I am inclined to think the vibratory theory will apply to these phenomena as it does to sound, and most probably to light.

By the time Faraday's letter was opened, Maxwell's mathematical theorems and Hertz's technological experiments had conclusively proved the existence of electro-magnetic waves. We are immersed in an ocean of frequencies, as Faraday predicted, surrounded by vibrations that are beyond our perceptual capacities.

***

At first Heinrich Hertz could see no practical purpose for his experiments. 'It's of no use whatsoever', he is reported to have said, 'this is just an experiment that proves Maestro Maxwell was right—we just have these mysterious electromagnetic waves that we cannot see with the naked eye. But they are there.' A little over a hundred years later, wireless technology—from radio's capacity to occupy the airwaves, to today's mobile phone, a device that has been adopted throughout the planet faster than any machine in history—has involved an ever more intimate engagement with Hertzian frequencies. Today, the 'mysterious waves' that surround us but that 'we cannot see' serve as a carrier signal for the millions of 'smart objects' increasingly embedded in all aspects of life.

***

The prehistory of wireless lies with the telegraph, a technology whose capacity to 'separate communication from transportation' enabled the sharing of messages at non-human speeds. By transmitting time across distance, the telegraph solved the problem of global simultaneity and, with the invention of the instant, spawned the modern world. By the early twentieth century, electric veins of transmission permeated the body of the earth, wrapping it in a single technologically generated, standardized time.

In 1967 global simultaneity fused with electromagnetic vibrations. It was then that the International Committee for Weights and Measures ceased to define the second as a micro-division of the seasonal year and established an atomic description, which tied the second to the rate of electromagnetic transitions in the hyperfine structure of the cesium-133 atom. From then on, time was determined by the designation for frequency: hertz (or cycles per second).

Today, the locative omnipresence of wireless devices ensures global synchrony through the GPS system, arguably the cell phone's killer app. GPS requires satellites with onboard atomic clocks accurate to within a billionth of a second. The drive for ever-greater temporal precision is vital for AI systems such as self-driving cars, automated weapons, and the growing field of augmented reality. With this, as William Gibson foresaw in *Spook Country*, comes the 'everting' of cyberspace such that the 'grid' now envelops the whole of the earth. Computation unfolds outward, escaping the limits of the machine, as the world crosses over to the other side of the screen.

***

Since the 1980s—the retrochronic date for the first generation of cellular systems—the ongoing transmission of wireless media has been concurrent with China's remark-able rise. While the telegraph was viewed as an alien invader, by the time electric communication went wireless, China was deeply embedded in the wave of techno-capitalist innovation that now ripples across the globe. The dramatic pace of this sociotechnological mutation reaches its fullest expression in the megacity of Shenzhen, on China's Southern coast, where most of the world's cell phones are produced. Shen-zhen has combined its role as factory to the world with a '*shanzhai*' street commerce, which emerged from the in-between cracks of the global economy. Plummeting prices combined with a distribution network of unprecedented scale and speed to create a hub of global electronic production. The markets clustered at Huaqiangbei overflow with wireless devices: phones, wearables, sensors and circuits. Cheap versions of the latest smart object flow from the rising metropolis, out to every corner of the planet.

***

The clock was invented in China, but it was not until centuries later, when it was rediscovered in Europe, that it transformed into what Lewis Mumford would call 'the key machine of the modern industrial age'. This cultural divergence in the fate of the mechanical clock is an example of the Needham question—the name that has crystal-ized around the still lingering problem of the ultimate compatibility between Chinese culture and modern technology. In his densely detailed tome *Science and Civilization in China*, Joseph Needham asked how a place famous for the four great inventions (*si da fa ming*)—compass, gunpowder, paper, and print—could have faltered in the modern period such that it was 'totally overtaken by the exponential rise in the West after the birth of modern science at the Renaissance'.[1]

In China, reaction to this 'great divergence' has been polarized. On one side is a determined resistance to 'foreign' technology (from the Boxers tearing up telegraph poles to the Great Firewall and the censorship of cyberspace). On the other is a belief, most forcefully expressed in the May Fourth Movement, but still prevalent in the internet politics of today, that in order to modernize, China must Westernize. Both tendencies are deeply entangled in the late Qing intellectual strategy, which mobilized the ancient cosmic dualism *ti/yong* (*essence/use*), in order to create a barrier that would separate Chinese cultural and intellectual heritage from the practices of a modern techno-scientific world.

***

Twentieth-century New Confucianism seeks to synthesize an abstract, reinvented tradition with an emergent modernity. Xiong Shili, one of the movement's founders, turned to Buddhism, and insisted on the nonseparation of *ti* and *yong*. The apparent division between a culture's essence and its practice, he argued, was a delusion born of attachment. 'Just as the water in the ocean is manifested as waves, *ti* is like the deep and still sea and *yong* like the continuous rise and fall of its many waves.'

---

1.   J. Needham, *The Grand Titration: Science and Society in East and West* [1969] (Abingdon and New York: Routledge, 2005), 285.

Xiong's most famous book *Xin Weishi Lun* (*A New Treatise on Consciousness-Only*) fuses Yogacara Buddhism with the continuous transformation described in China's most ancient classic, the *Yijing* (or *I Ching*). For Xiong, China's best hope in facing the alien world that had arrived on its shores was a return to the cosmo-ontology of the wave. Implicitly influenced by the ambient electric frequencies that were everywhere around him, Xiong saw in the 'uninterrupted flash upon flash of lightning' the ceaseless ancient pulsation of contraction (*xi*) and expansion (*pi*), generation and extinction, that is the essence, Ultimate Reality or Original Body of the Ten Thousand Things.

Xiong's disciple Mou Zongsan pushed the project of integrating Chinese tradition with modernity even further through an engagement with Immanuel Kant. Using the conceptual language of Chinese Buddhism, he translated all three of Kant's critiques. Mou's aim was to show that Chinese thought offered a path beyond the limits of reason, opening a gate to what Kant deemed impossible: intellectual intuition, or access to the thing-in-itself. Mou held that this practical, rather than theoretical knowledge was con-tained within the Confucian, Taoist, and Buddhist tradition. He likened the awakening they promised to a vibrational event. In describing these vibrations, Mou turns to a calendric event. In early spring, around the fifth of March, when the sun reaches 365 degrees, the Chinese *nongli* (or agriculture calendar) shifts into the third of its twenty-four solar periods. This marks the beginning of *jingzhe* (the awakening of the insects) when the buzzing reverberations signal nature's renewal and the wavelike reccurrence of time.

***

Both the Wifi Alliance, which aims to 'connect everyone and everything', and China's own GPS-like satellite system *Beidou*, chose as their symbol the undulating image of Yin/Yang. The logo—the most famous diagram of Chinese thought—resonates with the wireless wave at a variety of scales—from the techno-capitalist wax and wane of historical time to the microtemporal electromagnetic frequencies, through the myr-iad vibrations of our now ubiquitous mobile devices that grow ever more immersive, autonomous and smart. In tapping into the electromagnetic field that is everywhere around us, wireless media operates as the underlying abstract infrastructure of our cyborgian existence. Beyond their role as communication devices, these censors and circuits effectuate a 'cosmological revelation', trafficking in frequencies that we cannot directly perceive. This vibratory plane hosts a myriad of nonhuman sentient agents, which increasingly constitute the invisible, abstract, alien atmosphere that Faraday, long ago, secretly foresaw.

2014: THE VISUAL MICROPHONE

Lendl Barcelos

> O olhouvido ouvê
>
> (The eyear hearsees)
>
> — Décio Pignatari, *Teoria da Poesia Concreta*[1]

The visual microphone is an algorithmic process that transmutes pixel data from video recordings into sonic information. Funded in part by the Qatar Computing Research Institute and the American National Science Foundation, the research and its (patent-pending) results were made public in 2014 and brought together researchers from labs at MIT, Microsoft, and Adobe.[2] Whereas previous attempts to recover audio signals from visual sources required systems to project onto vibrating surfaces—an *active* intervention—the technique of the visual microphone suggests that minute surface vibrations contain enough information for audio to be *passively* reanimated without the need to intervene. Even with the sound turned off, a moving image betrays situated oscillations: a camera's visual register is détourned to become a medium for listening in. Dormant sound is reawakened and made audible. By 'turning visible everyday objects into potential microphones' the researchers have developed techniques to catalyze yawning ears toward the summons of spectral vibrations thought to have been silenced.[3] With the visual microphone, infra_perceptible sound events of the past, once recorded as video, can be brought back to life.

The process underlying the visual microphone consists of tracking small surface vibrations produced when sound collides with an object. It is not sound that is being recorded, but the surface effects it elicits. A camera captures these tiny movements and registers them as minute pixel fluctuations varying over time. Then the subtle motions are amplified—while maintaining their relative differences—and transposed within the

---

1.   Quoted in A.S. Bessa, 'Sound as Subject: Augusto de Campos's *Poetamenos*', in M. Perloff and C. Dworkin (eds.), *The Sound of Poetry, the Poetry of Sound* (Chicago: The University of Chicago Press, 2009), 219–36: 219.

2.   A. Davis et al., 'The Visual Microphone: Passive Recovery of Sound from Video', *ACM Transactions on Graphics* 33:4 (2014), 79:1–10.

3.   Ibid., 1.

bounds of the audible. In this way, the algorithmic process of the visual microphone can comb through videos in order to search for latent sounds unheard. Researchers initially tested this technique with high-speed cameras only, but later showed that similar results could be achieved using standard consumer cameras by compensating for a reduced frame rate. As a proof of concept, a sound system and a bag of potato chips (i.e., an object capable of being perturbed) were placed inside a soundproof room. A camera, set up outside of the room behind a pane of glass, registered the micro-movements of the surface of the bag of chips. Although the sound system remained out of frame and the camera was positioned in the adjacent room, the sound played in the room could be heard once the visual microphone had reanimated it—albeit in a degraded form.

What is striking about this technique is that it opens up the possibility of listening in on the acoustic space where the camera captured the footage—even when that space has since decayed to ruin. So long as there are objects within the camera's field of view able to reflect sound events, the (micro-)movement-image produced can be used to sound the depths of the area around the camera. A camera does not have to be directed toward the sound's source to register a trace of its perturbations. Given the increasing ubiquity of cameras—on mobile devices and as part of a paranoiac heightened surveillance—this means that, even when sounds occur beyond the sensitivity range of an audio microphone, it is now possible for them still to be brought within earshot. In a sense, Thomas Edison's attempt to hear the voices of the dead has now been transmuted into a synaesthetic programme whereby faint acoustic impressions can be visually amplified so that spectral voices and the spaces they once inhabited can be heard.

The technique of the visual microphone reveals that hidden within the visual lie ineffables awaiting articulation. Technologies continue to be developed that can register, transduce, amplify, and transmute phenomena at scales well below the thresholds of human sensitivity, opening up sensory fields as yet unsensed. With its capacity to open lines of communication between the vibrational space around the recording camera, its resultant (micro-)movement-image, and our ears, the visual microphone is just one example of these technologies. Here, our ears are transported back into past videos through variations of pixel data in order to reanimate sounds thought to have perished. We are left to wonder: What inaudibles lurk in the micro-movements of pixel fluctuations?

THE THING

Amy Ireland

In 1945, under instructions from the NKVD, Russian intelligence operatives planted a passive covert listening device—designed by Léon Theremin and concealed inside a carved, wooden replica of the Great Seal of the United States of America—in the American embassy in Moscow.[1] The bugged seal was gifted to the USA by members of the Vladimir Lenin All-Union Pioneer Youth Organization and, once installed, was sensitive to remote activation by means of a microwave signal transmitted in the vicinity of the embassy building. It was successfully used to record classified discussions taking place in the American ambassador's office between 1947 and 1952. For highly suggestive reasons—its retrospective recognition as an occult resonator chief among them—the device was dubbed 'the thing' by baffled employees of the United States Secret Service.

In a demonstration of exemplary camouflage tactics, the thing had lain hidden in plain sight for seven years, undetectable without contingent or clandestine knowledge of the hypersonic frequency it operated at, before it was discovered accidentally, while receiving its illuminating signal by a British radio operator monitoring Russian air force traffic in 1952. As it needed no power supply and contained no active electronic components, routine security sweeps had consistently failed to detect the thing's presence, but it was finally given up by the howl of positive feedback it emitted whilst active during a well-timed counter-surveillance sweep (using a tunable receiver operating at 1800 megahertz) ordered in response to the tip-off.[2] American officials removed

1.   The NKVD was the Russian Commissariat for State Security between 1931 and 1946. It is one of the predecessors of the KGB. Léon Theremin is the name adopted by Lev Sergeyvich Termen whilst touring the United Sates and Europe in the late 1920s.

2.   P. Wright, *Spycatcher: The Candid Autobiography of a Senior Intelligence Officer* (New York and London: Viking, 1987), 25.

the device and kept quiet, hoping that it might provide additional bargaining power in future negotiations with the USSR.

Significantly, discovery of the thing was not coincident with an understanding of how it worked, and it would take the American, British, and Dutch security services between eighteen months (in the case of the British) and fifteen years (for the combined Dutch/American effort) to reverse engineer and successfully replicate its mechanism. In yet another twist of serendipitous denomination attentive, perhaps, to the epistemological problem that accompanied it, the first working MI5 prototype was nicknamed 'Black Magic' by its chief engineer, Peter Wright. Indeed, Theremin had accomplished what seemed like a sorcerous act to anyone unfamiliar with the possibilities that electromagnetic radiation offered the world of espionage and clandestine communication. Theremin's brief was for a device that would function wirelessly, did not require traditional microphones, and could be introduced into the ambassador's residence without raising suspicion. Faced with the implicit threat of being returned to a Gulag camp in Kolyma, or executed, Theremin delivered. The thing comprised only a few simple components: a silver-plated copper cylinder roughly 13 millimetres deep and 20 millimetres in diameter, and an insulated rod antenna whose original length and corresponding resonant frequency is a matter of much contention among those who have attempted to reconstruct the device. Inside the cylinder, the rod terminated in an adjustable tuning post, forming a capacitor with a diaphragm that extended across the resonator's open face. Exact descriptions of the original bug have never been released, leading to speculation as to whether or not it was simply an early example of RFID technology, or a more complex device employing harmonic reradiation. In the case of the latter, the antenna would have acted as both the receiver for the illuminating signal and the device's transmitter. Wright's reconstruction for MI5, renamed SATYR and used by the British, American, Canadian, and Australian secret services, functioned with a resonant frequency of 1400 megahertz, although higher frequencies have been hypothesized for Theremin's original design.[3]

The enigma surrounding the thing was to have enduring repercussions in the paranoid psychoscape of Cold War diplomacy and state-sanctioned espionage. In 1962, ten years after dismantling Theremin's bug but only two years after public revelations of its

3.   I.W. Conrad, 'Internal FBI memorandum, 8 May 1953', <www.cryptomuseum.com/covert/bugs/thing/files/19530508_fbi.pdf>; G. Brooker and J. Gomez, 'Lev Termen's Great Seal Bug Analyzed', *IEEE Aerospace and Electronic Systems Magazine* 28:11 (2013), 4–11.

discovery, the United States became aware that their embassy in Moscow was once again the target of abnormal levels of electromagnetic radiation. Was this evidence that a new passive listening device had been installed in the building, or something more ominous? Several theories were advanced at the time, although none were ever officially confirmed as motivation for the bombardment. If it wasn't destined to illuminate a newer model of the thing, the electromagnetic signal could have been connected to Theremin's subsequent innovation, a method of microwave-based audio surveillance known as the 'Snowstorm' system.[4] Others considered it to be a bluff, designed to lead the Americans into believing that the Soviets possessed an unfathomable new piece of military technology. Some went even further and suggested that it was indeed evidence of the latter, hypothesizing a system that employed electromagnetic radiation to induce neurasthenia or other biological and behavioural changes in its subjects, including the use of 'synthetic telepathy', or mind control. This latter hypothesis inaugurated an experimental research programme overseen by ARPA at the Walter Reed Army Institute of Research in the late 1960s.[5] The programme, dedicated to exploring the biophysical effects of microwave radiation, is known in certain circles as the infamous Pandora Project.[6]

The thing can be counted among the catalytic moments of covert electromagnetic surveillance technology in the twentieth century. The existence of synthetic telepathy may remain a peripheral theory, but the contribution of Theremin's invention to the development of RFID tags and scanners, tracking devices, and ubiquitous telesurveil-lance, not to mention its legacy of ambient paranoia, carries over the threshold of the millennium, perpetuating itself imperceptibly in the deep structure of twenty-first-century control dynamics.

4.   A. Glinksy, *Theremin: Ether Music and Espionage* (Champaign, IL: University of Illinois Press, 2000), 260–61.

5.   D.R. Justesen, 'Microwaves and Behaviour', *American Psychologist* 30:3 (1975), 391–401.

6.   P. Brodeur, *Currents of Death* (New York: Simon and Schuster, 1989), 91–2; N. Steneck, *The Microwave Debate* (Boston, MA: MIT Press, 1984), 94–5.

LARGE HADRON COLLIDER: THE ULTIMATE UNDERGROUND GROOVE

Toby Heys

At the turn of twentieth century, the notion of propelling two forces backwards and forwards, to meet at the point where they negate each other, takes hold within military think tanks across Europe. They are searching for a more expansive comprehension of the collision of the subliminal message and the spoken word, matter and anti-matter, and existence and nonexistence. Research is carried out into these phenomena through war, entertainment systems, vortexes of urban unrest, mass anxiety, mobile music playback technologies, and the plethora of sonic portals that make up the global matrix of the speaker network. For it is the human sensorium that will become the next battleground to be mapped, strategized, and conquered. At the end of the tenth century of the second millennium, the military will brief government officials about their cultivation of hypogean activities and networks. Comprehending that the underground is the space in which technological innovation meets speculative thinking, the dark occultists of the white Western ring of scientific power will, in response, magnify AUDINT's 1946 TwoRing Table experiments[1] by technically coming full circle and completing the evolution of the underground groove.

It is 1998 at CERN, Switzerland. Construction begins one hundred meters below ground level in order to assemble the Large Hadron Collider, a massive locked groove measuring twenty-seven kilometres in circumference. In this particle-accelerated underworld the purest form of collision will be orchestrated by smashing particle beams

1.   In AUDINT, *Dead Record Office* (New York: Art in General, 2013), the TwoRing turntable first emerges as an idea when AUDINT members Walter Slepian and Bill Arnett meet with English cryptanalyst and mathematician Alan Turing, shortly after the end of WW2. 'They explore different ways of spinning recorded discs until Turing suggests the possibility of producing a table that has two arms, starting side by side at the 6 o'clock position on a static slab of shellac and working their way around a locked groove, tracing its curve until the needles collide at the 12 o'clock position. Turing's notion of playing a locked groove in its clockwise (future) state and anti-clockwise (past) state until the two arms turning around the static vinyl meet and collide with each other excites Arnett and Slepian.' (17).

of protons or lead nuclei together (six hundred million times per second) in an attempt to mimic the elemental circumstances of the universe in the first trillionth of a second that succeeded the big bang; this being the point at which everything we comprehend as having mass begins to be endowed with weight as the projected energy of the Higgs Field is attracted to fundamental particles.[2] Traversing science's circular answer to the cultural rise of ambiently organized violence, detectors trawl the subatomic wreckage after each impact. They record every percussive breakdown in an attempt to locate echoes of the celestial sticky voice that binds together the total mass of the universe.

Along with decoding divine voices, there are other notes on the agenda of the LHC score. Scientists will search for evidence of dark matter—the unlit and non-radiated dead quarter(s) of the universe. They will examine antimatter and the reason why the universe is not made up of it; supersymmetry, which predicts that each and every fundamental particle has an unperceivable heavier phantom twin, an ethereal operator adorned with a plethora of names such as the squark, the twin of the quark, the photino, the stuff of light; and finally and most crucially (for a certain select group of researchers) will be the investigation of extra dimensions whose existence they have to believe in, since otherwise their journey to the tenebrous side of enlightenment will have all been in vain. For the majority of scientists at the LHC, however, the ultimate goal is to make contact with the 'God Particle' (the Higgs Boson)[3] and thus with the forces that produced the known universe and our place within it. Once they have achieved this, their hopes rest upon the idea that they will be able to influence and modulate the resonant frequencies of all matter.

Fourteen months on from the initial magnetic quench problem that results in fifty superconducting magnets being damaged, the Large Hadron Collider reopens in November 2009 for both godly and ungodly business. Whilst the godly research is undertaken in the glare of the media spotlight, a camouflaged cabal of technologists, engineers, military personnel, and economists will quietly conduct another investigative program for a different entity—that of the god-damned. They will search for the ultimate force carrier and elementary particle that will provide answers as to how execrable, destructive, and malevolent phenomena are formed and propagated. They will listen for,

---

2.   For a concise explanation of the Higgs Field, see <http://physics.about.com/od/quantumphysics/f/HiggsField.htm>.

3.   In the Standard Model of particle physics, the Higgs boson is a recently confirmed elementary particle that was believed to exist (by Peter Higgs) since the 1960s. Its verification resolves many longstanding problems within the field of particle physics, such as why some fundamental particles have mass when, according to their interactions, they should have none.

and record, the compositional elements of dark matter. But it is here they will deviate from the mandated programme as they circumvent the cosmic tendons that exert their force upon galaxies (in the theorized form of the neutralino) and aim instead to find the Diabolus Particle. It is ventured that the accelerated collision revealing this obsidian speck will subsequently open up the vaults of the Bank of Hell; the only cadaverous fiscal institution European governments can turn to after 2012's economic cataclysm.

As tones of civil unrest come to score the post-9/11 period, culminating in the crises of Brexit and the possible implosion of the Euro (as well as the ushering in of the spectres of war that inevitably become palpable in times of severe economic downturn), the treasury of Erebus[4] offers a desperate final chance for economic absolution. As the fixers, programmers, and dealmakers must know, any deal between the Bank of Hell[5] and the Military-Banking complex will come at a price that can never be repaid. For if the living transgress the realm of the here and now and enter the databases of the dead, the present will lose its meaning. It will temporally fragment and decompose into the future and the past, meaning that humans will live through the end of linear time. Working day and night to unearth this particle, what the scientists do not know (but what AUDINT suspect) is that upon locating the Diabolus Particle in the underground groove, a ringing sound will be triggered; a clarion call for the gates of the Bank of Hell to be opened. A high-pitched set of heterodyning tones that will cause tinnitus in each and every human on earth—a global ringing that will allow a third-eared network to grow and transmit.

The perception of a high-pitched whining or whistling sound (with no apparent external source) will cause a rapid increase in the size of the *corpus callosum*, the great commissure responsible for connecting the right and left hemisphere of the brain (a neurological process that takes years to occur and is only evident in the brains of musicians).[6] It will also render each human being with a new embodied perfect pitch, which means that the *planum temporale*[7] will also become asymmetrically enlarged. Being bestowed with this capacity will allow humans to develop the efficacy of the

---

4.   In Greek mythology, Erebus, symbolizing dark and shadowy agency, was a primordial deity born out of the chaos of the void.

5.   The most significant economic institute of the undead, the Bank of Hell (translated as the 'underworld court' in Chinese) is the gateway administration, where the souls of the dead are initially judged by Yan Wang (the Lord of the Earthly Court). Within the court, hell money is used to leverage the standing of the soul.

6.   For further reading on the visualisation of musician's brains, see O. Sacks, *Musicophilia: Tales of Music and the Brain* (New York and Toronto: Alfred A. Knopf, 2007), 94.

7.   A part of the auditory cortex.

third ear and the ability to communicate through their entire corporeal structure as a transmitter/receiver. As such, the body will be able to perceive frequencies beyond the 20–20,000Hz vibrational straitjacket that the current somatic model finds itself bound within. In other words, the release of knowledge concerning the origins of the big bang and the release of the dead into the world of the living will give birth to a new sonic and somatic modality—that of the big ring. A new post-hearing world, in which the body becomes the eardrum.

RWD

RETURNS, REPEATS, AND LOOPS

DUPPY CONQUERORS, ROLLING CALVES, AND FLIGHTS TO ZION

Julian Henriques

In Jamaica, a duppy is a spirit or ghost of a dead person. They are undead, but unlike their cousins from the nearby Caribbean island of Haiti, the zombies, they maintain individual agency. Duppies usually take human form, though their feet are said to point backwards, in order to confuse anyone trying to track their footprints. They come out at night and are said to congregate under cottonwood trees. In Bob Marley's *Duppy Conqueror* the proverbial hero fights back against these ghosts—of his vanquished enemies perhaps?—and 'bullbucka' (bullies):

> Yes mi friend, me der 'pon street again [...]
> So if you a bullbucka, let me tell you this
> I'm a duppy conqueror, conqueror...[1]

Not surprisingly, the duppy has also been a popular figure in novels and poems as well as in song.[2]

The mingling of spirit and human worlds enriches us immeasurably. It allows the past to be alive, not dead to us, as is the case for the deadweight of technology with which Western culture burdens itself. But such (digital) technologies have also contributed to this intermingling by making instantly available the entire history of recorded music. In being flattening out and dehistoricised, the past has all become equally present, which appears to have whetted an appetite for the ancestors—a newfound love that appears to resuscitate and extend the lifespan of what might otherwise be considered dead media—retro analogue synths, cassette tape, and vinyl records.[3]

1. Bob Marley, *Burning* (Tuff Gong/Island, 1973).
2. From the classic ghost story by Herbert G. de Lisser, *The White Witch of Rose Hall* [1928] (London: Macmillan Caribbean, 1982) to Ferdinand Dennis's *Duppy Conqueror* (London: Flamingo, 1998).
3. This is literally the case with, for example, the 2016 Mercury Award-winning young saxophonist and band leader

Duppies are liminal figures caught 'betwixt and between' worlds, intertwining future and past, living and undead. 'The future is always here in the past', as Amari Baraka puts it.[4] The dead are always here amongst us, the living. This is the case for innumerable cultures, though not in that of the dominant Western materialist one, with the exception, perhaps, of the Gothic tradition. The particular past-to-be-future discussed in Jamaica is that of the ancestors of West African tradition as they express themselves in folkloric beliefs and proverbs, as with for example, 'Ev'ry cave-'ole 'av' 'im own duppy' (every cave has its own ghosts, or every family or person have/has their own problems).[5]

A rolling calf is a duppy not in human form but in the form of a raging bull with fiery eyes and flames flaring from its nostrils.[6] These animals are also credited with having a distinctive and equally terrifying sonic signature: the clanking of chains dragged from around their necks, and the bellowing noise they make and which causes them to be known also as roaring calves. I might have heard one myself. There lies a particular country graveyard, typically without a church, in a lush green valley in the Portland foothills of Jamaica's Blue Mountains, near a village that goes by the name of Nonesuch. One night I heard such a sound, or maybe an unsound—or perhaps it was a donkey's braying carried by the wind mixed with the tune from a sound system playing out from across the next valley.

On another occasion, to the tune of tree frogs, over supper, I was told by a palliative care doctor that it was commonplace for patients to know exactly the hour of their death, often with a sense of calmness and acceptance. This came, the doctor said, once the patient had been visited by a spirit or duppy of someone already departed, welcoming them to the afterlife. One patient described in exact detail the character of the duppy and the clothes it was wearing. The duppy turned out to be no one they had ever met, but the occupant of her same hospital bed, who had died some two weeks previously.

Shabaka Hutchins in all of his combos—Sons of Kemet (Kemet being the name for ancient Egypt; *Burn* [Republic of Music, 2016], *Lest We Forget* [Republic of Music, 2016]), The Comet is Coming (*Channel the Spirits* [The Leaf Label, 2016]), and as Shabaka and the Ancestors (*Wisdom of the Elders* [Brownswood Recordings, 2016]).

4.   A. Baraka, 'Jazzmen: Diz & Sun Ra', *African American Review* 29:2 (1995), 255.

5.   G. Llewelln Watson, *Jamaica Sayings with Notes on Folklore, Aesthetics and Social Control* (Gainesville, FL: University of Florida Press, 1991), 189.

6.   There are also other kinds of duppy such as 'long-bubby [breasted] Susan', 'whooping boy' riding 'three foot horse', and 'Old Hige'. See Anon [1904], 'Folklore of the Negroes of Jamaica', *Folklore* 15:1 (25 March 1907), 87–94.

But the undead can also threaten the living. Duppies are troubled spirits, malicious souls, who do not rest in peace at all and are feared as objects of *dread*.[7] They are said to cause accidents, make you lose money or love, and can even attack with a weapon; hence the need to conquer them. For Early B's *Ghostbusters* album cover, the influential but little recognised Jamaican graphic artist Wilfred Limonious captures this well known trope perfectly.[8] According to Jamaican folklore, to prevent the undead rising up, they have to be buried in the proper manner. The body has to be 'planted down' in the coffin by 'throwing a shovel full of parched peas into the grave. So long as they do not grow, the duppy cannot escape', as one account has it. 'A shrub planted in the grave upside down, that is roots out, is also efficacious'.[9]

Duppies don't so much lie between the living and the dead as *fly* between these two worlds. 'One bright morning when my work is over, Man will fly away home...', in the words of the traditional Revivalist song. The flight path for these spirits to escape the downpression of Babylon is set for the motherland of Zion, 'I say fly away home to Zion...', as Marley sings in *Rastaman Chant*, or, in *Duppy Conqueror*, 'Don't try to cut me off on this bridge now / I've got to reach Mt. Zion."[10] Traditionally, what prevents take off for Zion is the weighing down of the body caused by eating salt, salted fish and meat being part of the plantation owners' diet for their slaves.[11]

These flying ghosts have had considerable influence, not least in helping us understand how and why the island of Jamaica has become such a musical and spiritual powerhouse since the middle of the last century. Alexander Bedward (1859–1930) founded the Jamaica Native Baptist Free Church in August Town.[12] He convinced his followers that, rather than die a normal death, he would fly directly up to heaven as the Biblical prophets had been said to do. At the appointed time, it is said, Bedward

---

7.   Dread as in Jamaican lings: see J. Henriques, 'Dread Bodies: Doubles, Echoes and the Skins of Sound', *Small Axe* 44 (2014): 191–201.

8.   C. Bateman, *In Fine Style: The Dancehall Art of Wilfred Limonious* (London: One Love Books, 2016).

9.   M. Leach, 'Jamaican Duppy Lore', *The Journal of American Folklore* 74:293 (July–Sept. 1961), 207–15: 211.

10.   *Rastaman Chant*, also on Bob Marley's *Burning* (Tuff Gong/Island, 1973). The Zionites and their dub music have a similar quasi-redemptive role for the protagonist Case in William Gibson's classic cyberpunk novel *Neuromancer*. See an excellent account of this in L. Chude-Sokei, *The Sound of Culture: Diaspora and Black Technopoetics* (Middletown: Wesleyan University Press, 2015), 148–65.

11.   This trope of salt preventing the spirit's flight back to Africa is also taken up in Derek Walcott's epic poem *Omeros*, D. Walcott, *Omeros* (London: Faber and Faber, 1990).

12.   See Kai Miller's novel *Augustown* (London: Weidefield and Nicholson, 2016). The actual August Town community, dubbed New Jerusalem by the Bedwardites, abuts the University of the West Indies' Mona campus and has long been the location of Sizzla Kolanji's HQ, appropriately named Judgement Yard.

climbed up into an ackee tree, or, according to other versions, a tall building, to be borne aloft. But, as Prince Far I's lyric rather bluntly puts it in *Bedward the Flying Preacher*, 'Guess what happen? Bedward jump off the building top and break his neck'. Bedward was arrested and imprisoned before being carted off to a psychiatric hospital where he died.[13] Nevertheless, in the early years of the last century Bedwardism became one of the major antecedents for Rastafarianism, via Marcus Garvey and Leonard Howells. Together with Paul Bogle, leader of the 1865 Morant Bay slave rebellion, these figures inspired reggae music and lyrics that for more than half a century have rallied those striving for 'betterment' the world over.

We can also speculate that it is this same aeronautical theme that can be identified as the major trope of Afrofuturism, where it became famously astronautical, and the motherland was re-engineered as the *mothership*. Furthermore, the idea of flying between continents extends not only to interplanetary and intergalactic outer space, but also to the doubling *inner* space of dub music,[14] Burning Spear's album *Garvey's Ghost* being a good example, alive as it is with the echoic hauntings of the undead, tales and tails of sounds past. [15]

13.   Prince Far I, *Singers & Players, Staggering Heights* (On-U Sound, 1983).

14.   J. Henriques, *Sonic Bodies: Reggae Sound Systems, Performance Techniques and Ways of Knowing* (London: Continuum 2011).

15.   Burning Spear, *Garvey's Ghost* (Island Records, 1976).

DAYS OF WRATH

Eugene Thacker

Music has an intimate relation to death. Existing in time, music testifies to the melancholy brevity of existence; music is in fact this ephemeral, transient quality of everything that exists. E.M. Cioran writes: 'I know no other music than that of tears.'[1]

At the same time, we also know that music never ceases, even when the music's over. There is something in music that also resists time and temporality, that flails itself against the brevity of existence, its sound waves stretching out across the finitude of our hearing. Often music is composed of words, and yet the words the music expresses often transcend them, turning against the words, mutating them into something non-linguistic and yet communicable. It is no wonder music is often tied to ritual, the sacred, and the divine. But even this wanes. Music subsists in memory, often resurfacing, like a refrain, at the most unexpected moments—before again fading away into oblivion. Cioran again: 'Music is everything. God himself is nothing more than an acoustic hallucination.'[2]

But if God is an acoustic hallucination, then what of the Devil? The Devil's music is, of course, heavy metal. Should we then say that the Devil is not the smooth veneer of an 'acoustic hallucination', but the disharmony of feedback and noise? It has become a truism that Satanism operates on a logic of inversion, and this has undoubtedly influenced the way we culturally view harmony and disharmony, consonance and dissonance, signal and noise. The Satanic Black Mass, for instance, inverts the Catholic Mass nearly point for point (the inverted cross, the desecration of the Host, and so on). Given the import of the motifs of divine light and divine life in the traditional Catholic Mass, it

---

1. E.M. Cioran, *Tears and Saints*, tr. I. Zarifopol-Johnston (Chicago: University of Chicago Press, 1995), 11.
2. Ibid., 54.

would seem that the pinnacle (or nadir) of the Black Mass would be the inversion of divine light and divine life—an affirmation of demonic darkness and death.

If this is the case, then what does one do with the Requiem Mass (*Missa pro defunctis*), the Mass that in fact commemorates, and even celebrates, death? To simply invert this into a 'Mass for Life' would be tantamount to affirming the traditional Mass itself. In a sense, to negate the Christian Mass is all too easy, since the motifs are laid bare in their dualism—sanctity, chastity, transcendence, light, beatitude, and the afterlife. One has simply to systematically invert them via a kind of demonic algorithm. The problem, then, is the way in which opposition itself frames both the Catholic Mass and the Black Mass—life vs. death, divine vs. demonic, form vs. chaos, harmony vs. cacophony.

However, a look at the development of Western sacred music reveals numerous elements in early and medieval Christianity that would make even the most devout attendee of the Black Mass jealous: resurrection and the living dead, cannibalism and vampirism, corporeal metamorphosis, demonic possession, and a sophisticated poetics of eschatology.

In a sense, the Requiem Mass already is an inversion of the traditional Mass, full of ambiguities, spiritual crises, and a world rendered as sorrow and despair. The Requiem is already a Black Mass. Ostensibly a religious rite memorializing the dead as they pass on to the afterlife, the Requiem is unique in the repertoire of Western sacred music, in that it is an extended musical meditation on death, finitude, and—as we shall see—on the horror of life itself.

The Requiem occupies a special place in the sacred music tradition in the West. As a central part of Christian ritual, the traditional Mass is dedicated to the affirmation of the divine; as a Mass for the Dead, however, the Requiem is also an evocation of a whole host of apocalyptic elements, from the images of the *Dies iræ* (Day of Wrath), to corpses turning to ash, to warnings of evil spirits and 'demonic reports'. If the Requiem is a celebration, it would appear to be a celebration of death—or, more accurately, an affirmation of the life-after-life that death signifies in the apocalyptic tradition. While Requiems were composed throughout the classical, Romantic, and modern periods, it is in the emergence of the Requiem itself as a musical form that one witnesses the basic dichotomy that defines the Requiem—a celebration of negation, the exuberance of the void, the life-affirming ritual of death.

REVIVAL

Agnès Gayraud

A revival is the renewal, in a given culture, of one of its former practices or expressive forms that has fallen into disuse. The renewal implies a conscious effort to reproduce this older practice, to adhere to its genuine specificities. It emphasises fidelity to the original features of the revived art or practice over the newcomers' capacity for innovation. Once the practice or the expressive form has been considered as lost and worthy of being restored to life, its revival would be the reinjection of a piece of past culture into the present culture, like a transplant of a dead part of culture into its living body as a way to revive this very body. At some point, revival quarrels with the present situation of culture and art. Within culture, it turns against culture's historical process of consigning the obsolete to oblivion, and challenges the legitimacy of a society's present state for having buried some of its former practices. Driven by a nostalgia for a supposedly genuine popular art, the revivalist believes in the superiority of *roots*, ethnic or religious communities rather than nations, nations and their *Volksgeist* rather than states or superstates.

When did the word 'revival' appear in modern culture? According to Adorno, the modern occidental impulse for 'revivals' dates back to the eighteenth century, and the premises of the culture industry.[1] In Europe, as the first musical mass successes occurred, in the field of the *Deutsches Singspiel* and of the Italian *opera buffa*, regret for the passing of an immaculate popular art was already emerging, bound to small and mostly unindustrialised communities. As a compensation for this *loss*, a few decades later, the German Romantics von Arnim, Brentano, and the brothers Grimm collected popular songs from old, remote Germany, which the latter published in their famous volume of *Children's and Household Tales*, and the former in the collection *Des Knaben Wunderhorn*.

1.   T. Adorno, *Current of Music: Elements of a Radio Theory* (Frankfurt am Main: Surkhamp, 2006), 432.

Reviving these old stories for children—as an analogon of a fantasised childhood of humanity—they were reacting explicitly to the experience of disenchantment and death within culture. The *volkisch* poetics they intended to preserve (while rearranging it for the sake of contemporary urban readers) thus provided a compensational object. Ironically, though, their romantic dream to save the lost treasures of popular poetry and to hand them down to the 'people', their ghostly depository, had also attested to the disappearance of the living experience of popular poetry and to a reification which, far from saving it, signalled its irreversible end. Isolating art forms from life forms, their romantic revival seemed condemned to testify to the definitive death of what it wanted to revive.

But there are different ways to *revive* something. The mediation introduced by recording techniques caused a shift, for instance when folklorists like John and Alan Lomax, at the beginning of the twentieth century, were able not only to recall former musical and vocal practices through written archives and documents, but to record them in remote places of the country supposedly untouched by the encroachments of industrialised society. In the 1920s, if the US 'old times music' recordings were obviously still acting as a cultural compensation for the industrialisation of culture itself, the recorded voices of peasants, prisoners, or hillbillies from the Appalachian Mountains produced something more than a renewal of the past: under cover of geographic remoteness, it *resurrected* American pioneers' past (the geographic remoteness tending to also symbolically evoke a remote time), or at least shaped the fantasy of this past, seen to have been suddenly brought back to life. With their sound filled with scratches, deterritorialised and reproducible, the original field recordings reincarnated a remote singer's voice and instrument within the present experience of the listener. In this fashion, the magic of revival could operate in a far more convincing way: as a direct transplant into the present of the (synthesized) surviving flesh of the past. As a matter of fact, this ability to perform an audio 'resurrection' on demand has been essential to the possibilities of recording technology. The globalized broadcasting of recorded popular music has filled the world with such revivable voices for more than a century now.

With the internet's contemporary hypermnesia, recordings of dead voices probably speak louder now than living ones. Nonetheless, this has more to do with the problematics of the accumulation of the archives than with the original impulse to revival, which is, precisely, a protestation against the reduction of cultural heritage to mere archives. Indeed, a revival may never be fully performed by (even 'revivable') archives alone; it also wants to recall an actual way of life, in other words, a form of life. In his *Chronicles*, Bob Dylan mentions the fact that, during the concerts he used to organise at his loft on Third Avenue in New York during the early sixties, Alan Lomax made it a point of honour to take black prisoners out of their prisons to come and sing field hollers.[2] The 'spiritual experience' that Dylan recalls has arguably a lot to do with this Holy Grail of an archetypal musical expression directly bound to a rural life form, here (hopefully) genuinely incarnated by a black prisoner for the appreciation of white urban intellectuals. The fact that these were live events, attended by a small community of avant-garde amateurs, preserved at least the illusion that the observers did not, so to speak, kill the dodo they were observing. But their aesthetic (if not political) urban community couldn't but confirm the tragic auto-betrayal of any revivalist temptation: the dislocation of the art from its life form. This is probably why the revivalist is often a fetishist, a collector: George Lucas's *American Graffiti* (1974) wouldn't be a nostalgic movie of the youth culture of the end of the fifties without its collection of cars (Chevrolet Impala, cabriolet Thunderbird) and sweaters. But to avoid the fate of the entomologist—once a nature lover, he ends manipulating dead skins and bones—a consequent revivalist must drive the car daily, and wear no other sweaters.

Ultimately, the dialectics of revival without devitalization prove to be very tricky. Joe Boyd described for instance the ambiguous process of the revival of Bulgarian folklore, from the foundation of the 'Soviet Ensemble' by ethnomusicologist and composer Philip Koutev up to 4AD's *Le Mystère des Voix Bulgares* compilations in the nineties. On the one hand, 'validating ancient peasant culture, treating it as worthy and exciting and sexy and true' not only brought to public attention previously unheard music but also, during the decades of Russian domination, served as a valuable 'Soviet anti-matter'.[3] On the other hand, transforming Bulgarian traditional singing into a 'picturesque' industrialized commodity, prized for its (at the time, still genuine) pre-industrialized, pre-globalized look and sound, could only produce the same old effect: a kitsch residue of a fantasised rural past. But as Boyd notes, there is still an 'intensity and a freshness' there, which actually owes to 'the conflicts that still marked the intersection of politics and culture in Eastern Europe'. This is maybe the only way for a revival to avoid pure postmodern contingency: when the survival of a people is at stake.

---

2.   B. Dylan, *Chronicles, Volume 1* (New York: Simon & Schuster, 2004), 71.

3.   J. Boyd, 'How Stalin Invented World Music or Le Mystère des Voix Soviétiques', <http://oook.info/musics/stalinworld. html>, first published in *The Independent*, 18 May 2003. See also J. Boyd, *White Bicycles: Making Music in the 1960s* (London: Serpent's Tail, 2006), chapter 3: 'Hearing traditional musicians when they first emerge from their own communities is a wonderful experience but impossible to repeat: the music is inevitably altered by the process of "discovery".'

HOLOJAX

Toby Heys

The year is 2030 and, as much as the numen of Hades[1] has influenced and shaped the topography of holographic culture, it is in fact the Erotes—the winged retinue of Aphrodite[2]—that initially direct its business. To this end, a venereal system coming out of 'Electronics Avenue' (Zhongguancun) in China instantly captivates a holocore generation and, in doing so, cracks the market wide open. Shorn of the lumpen wavefront hardware necessary for large commercial venues, this new system has been miniaturised for household operation. For regular users, this means that they can set up on a table or shelf and project the musical dead into their front rooms, interact with them, and more. And so, the initial business-driven scheme—to stadium tour the dead—has entered a Rabelaisian reversal, as the departed are habituated into domesticated routines and regulated patterns of behaviour; a servile regimen that they must have tenaciously striven to avoid in their previous animate incarnations.

And yet here they are, at the behest of your voice and only a swipe of a finger away. Ask your selected entertainer to play a song, an album, or a mix and they comply with starlit élan. Learning as it goes, the device runs what is basically a pimped-up artificial neural network from the fifties—the perceptron algorithm[3]—to predict not only the choice of track but also the grade of virtuosity displayed by the user. Move on up to Level 2 and there is a choice of karaoked collaborations. Level 3 notches up the complexity of the interaction by offering jam sessions that the buyer is expected to instrumentally partner on. Level 4, however, is only available on one model, which although officially named 'Pothos', is known on the street as Holojax. And it is the

---

1. A reference to the divine will of the underworld in Greek mythology.

2. In Greek mythology this group of winged gods were equated with sex and love.

3. For further insight into the biologically inspired perceptron algorithm, see H. Daumé III, 'A Course in Machine Learning', version 0.8, August 2012, <http://www.ciml.info/dl/v0_8/ciml-v0_8-ch03.pdf>.

Holojax, when ordered through the underground beige market, that offers the sexual options, a beguiling range of projected pleasures.

Sporadically enjoyed in small groups, Holojax is more regularly fired up by individuals, the physical intimacy of the experience deregulating and uncoupling the kind of prosaic sexual relations that had previously been customary for blood-driven partners for hundreds of thousands of years. While still frowned upon by older generations, for those not yet in need of an epigenetic reset (not old enough to be in danger of being affected by harmful genetic markers relayed by previous relatives) it is the after-hours comedown of choice. Or at least this is how things started out. A year after it was made publicly available, myths were already circulating—of users being hooked up to the machines for inordinate durations, while they are fed and changed by employees and personal assistants, who have become known as 'watchers'. In changing the rhythmical location of interactive holography to the bedroom, Holojax facilitated a new kind of holographic necromancy. In terms of the sales pitch, it is an easy one: fucking the dead as the ultimate form of home entertainment.

To get started, all one needs is a modded suit, preferably one of a more pliant nature than is on offer from Tokyo's twin-tech prefecture, Akihabara—similarly referred to as 'Electric Town'. Full of actuators, protrusions, lube portals, and heat transfer sensors, once such a suit is plugged in, the pleasure can last for hours, days, or as long as you can pay for the Holohi to last. The only real limitation in this scenario is that of arterial fortitude—how many chems the somatic system can take and how effectively they are distributed. The Holohi itself comes from the intake of a mixture of the synthetic drug DB4 and the hormone oxytocin. Officially packaged as DBX, it is sold on the street as 'Ox'. The liquid drops are gently fed into the ears via a tiny tube that runs into the canal and delivers the stimulant onto the cochlea, where it forges its own pathway into the auditory nerve, creating a full-blown multimodal 'enhancement' in the user.

With advancements in light modulation techniques, information encoding, and computational power, things have come a long way since the days of the rapparitions.[4] 3D acoustic manipulation, the process that allows particles to be moved by acoustic waves, has become commonplace.[5] So much so that the music played by the holo musician no longer simply lets you dream, it literally takes you out of your head.

4. Dead rappers digitally exhumed as holographic entities in the early 2010s.
5. For early research in the area, see Y. Ochiai, T. Hoshi, and J. Rekimoto, 'Three-Dimensional Mid-Air Acoustic Manipulation by Ultrasonic Phased Arrays', *PLOS ONE* 9:5 (2014): e97590, <https://doi.org/10.1371/journal.pone.0097590>.

By adding a new modality to the sense of touch, the technology coaxes the player to envelop and penetrate the materiality of the music. Given this new visceral capacity, it becomes readily apparent why Holojax has become a global phenomenon, as music becomes corporeal and takes the form of a lover, a fuck buddy, a victim.

DIGITAL IMMORTALITY

Julian Henriques

The idea of digital immortality is not new. The word *digital* has remained the moniker for 'the latest technology' for three decades now.[1] We are technophiliacs because, as Freud might tell us, besides our own shit, technology is the one thing we make ourselves. Humankind—men in particular—have always tended to fall in love with their creations. This has been the case from the Greek myth of Pygmalion's most beautiful ivory statue, to the marvel—again scatological—of Jacques de Vaucanson's defecating mechanical duck of 1739.[2] This perhaps was the inspiration for Julien Offray de La Mettrie's bold proposition *Man a Machine*, published in 1748.[3] The philosophical claim that we are ourselves actually only machines was of course made by Rene Descartes almost exactly a century earlier, in 1637.

The idea of digital immortality holds the promise of every new technology. It will solve all our problems, even the major problem of life—that is, death. The Russian scientist Nikolai Fedorov, who inspired the Soviet space programme, submitted that the true ambition of science should be to raise the dead.[4] Even the humble gramophone was first advertised as a channel for communications beyond the grave.[5] According to Yuval Noah Harari's latest bestseller *Homo Deus: A Brief History of Tomorrow*, technology

---

1.   That is, if we date its impact from the adoption of the MIDI interface in the early '80s. The World Wide Web did not become publicly available until 1991.

2.   Jacques de Vaucanson (1709–1782), a French inventor and builder of automata, exhibited the duck at the Académie des Sciences, Paris.

3.   J. Offray de La Mettrie, *Man a Machine: and, Man a Plant* [1747 and 1748], tr. R.A Watson and M. Rybalka (Indianapolis: Hackett, 1994). Fritz Kahn (1888–1968) continued this tradition with his popular science diagrams: see U. von Debschitz and T. von Debschitz, *Fritz Kahn: Man Machine = Maschine Mensch* (Vienna: Springer, 2009).

4.   See J. Gray, *Straw Dogs: Thoughts on Humans and Other Animals* (London: Granta Books, 2002), 137–8, <http://turingchurch.com/2015/09/28/technological-resurrection-concepts-from-fedorov-to-quantum-archaeology/>.

5.   See J. Henriques, *Sonic Bodies: Reggae Sound Systems, Performance Techniques and Ways of Knowing* (London: Continuum, 2011), 190.

will make (some of) us gods.[6] Prometheus eat your heart out—or rather have your liver eaten out of you, which of course was precisely his fate for stealing one of their gifts.

This short text examines the key assumption upon which the idea of digital immortality can be said to rest, or maybe laid to rest. This is the Cartesian rationalist orthodoxy, which defines us only as minds—as cranial operating systems that can subsequently be separated from bodies. This idea is both supremely rationalistic and, in contradictory fashion, vauntingly hubristic; simultaneously both highly gendered and disembodied. It presents itself as the pinnacle of rationalistic scientific progress, while at the same time appealing to ancient superstitions and basic instincts. As Freud wrote: 'in the unconscious every one of us is convinced of his immortality'.[7] The idea of digital immortality fits well with current trends in so-called transhumanist and posthumanist thinking.[8] It fits even better with the techno-fetishist—if not techno-fascist—business plans of digital corporations.

The digital undead are already among us. What can be called first stage digital immortality is currently on offer—the grandiose projection of the startup Eternime being a case in point: 'Become virtually immortal [...] We want to preserve for eternity the memories, ideas, creations and stories of billions of people'.[9] A person's pattern of life on the net, every click, text, message, exchange, and purchase preserved in a corporate database, not to mention innumerable uploaded Instagram images and memorial Facebook pages.

Such a digital archive, though more comprehensive, is essentially no different from conventional analogue immortality, as in a collection of books, images, diaries, etc. But the claims of Eternime go further: the company 'creates an intelligent avatar that looks like you. This avatar will live forever and allow other people in the future to access your memories'.[10] We can imagine these avatars populating the soulless necropolis of stage one digital immortality, perhaps long after human extinction.

Second stage digital immortality is more ambitiously conceived as some sort of upload of our 'minds' into a computer software. Currently the Digital Immortality Institute offers 'social networking between the living and the dead',[11] and the world's

---

6.   Y.N. Harari, *Homo Deus: A Brief History of Tomorrow* (London: Harvill Secker, 2016).

7.   S. Freud, *Reflections on War and Death*, tr. AA. Brill and A.B. Kutter (New York: Moffat, Yard and Company, 1918), 62.

8.   See F. Farrando, 'Posthumanism, Transhumanism, Antihumanism, Metahumanism, and New Materialisms: Differences and Relations', *Existenz* 8:2 (Fall 2013), 26–32, <http://www.existenz.us/volumes/Vol.8-2 Ferrando.pdf>.

9.   <http://eterni.me/>.

10.   Ibid.

11.   <http://www.digital-immortality.org/>.

best known scientist Stephen Hawking is quoted as saying 'it's theoretically possible to copy the brain onto a computer and so provide a form of life after death'.[12] Google's Raymond Kurzweil predicts the so-called singularity, when machines will have become smarter than humans.[13] All the better, then, to use these machines to escape the biological limitations of our mortal coil in favour of a silicon base, even though this is a house that is, literally, built on sand.

The immortal soul and the idea of the uploadable mind are different, in that the mind needs a material—or digital—repository, whereas the soul does not. Historically the indestructible soul was born at the instant we could anticipate our inevitable demise. This momentous realization is said to have occurred in Orphic cults, from whose mystical ideas Pythagorean philosophy emerged. In the Eastern traditions this soul travels the circle of reincarnation. In the West it had an entire afterlife to inhabit, as in Ancient Egyptian religion and the Christian Kingdom of Heaven.

What the soul and the mind share is a common fear of the flesh. The soul has to be saved from the seductive pleasures of incarnation, vividly depicted, for example, in Hieronymus Bosch's 1510 painting *The Garden of Earthly Delights*. This saving of souls was the job of the Church.[14] Today the idea of digital immortality assuages our fear of having to live with the consequences of bio- and climate catastrophe. Abandoning the natural biological world in favour of a silicon one is touted as the next stage in 'human' evolution. This is the job of the digital technology corporations.

The memories that our minds require have traditionally been provided by cultural artefacts.[15] The fragilities of the digital domain are little noticed, given our obsessions with digital cloning and ubiquitous access to instant information. But without a completely reliable storage system, any idea of immortality is likely to be as short-lived as that of cryogenics, not least because both require a continuous electrical supply. In fact, the *older* the technology, the greater its reliability.[16] The cave paintings of Lascaux, for example, have lasted twenty thousand years; papyrus scrolls in clay pots have done well too, even two-inch analogue recording tape is more reliable than any hard drive.

---

12.   'Stephen Hawking Predicts "Digital Immortality" For The Human Race', *Medical Daily*, 21 September 2013, see also <http://www.thespacereview.com/article/873/1>.

13.   R. Kurzweil, *The Singularity is Near* (London: Duckworth & Co, 2006); see also Singularity University, <http://singularityu.org/>, and N. Bostrom, *Superintelligence: Paths, Dangers, Strategies* (Oxford: Oxford University Press, 2014).

14.   See M. Foucault, *The Care of the Self: Volume 3 of History of Sexuality*, tr. R. Hurley (New York: Vintage Books, 1988).

15.   P. Connerton, *How Societies Remember* (Cambridge: Cambridge University Press, 1989). Connerton draws on Maurice Halbwachs's concept of collective memory.

16.   J. Henriques, 'Thinking without Trace', *Visual Culture* 1:3 (2002), 355–8.

Indeed, oral histories are found to have sustained themselves for seven thousand years, according to researchers of Australian aboriginal traditions.[17]

By losing our bodies to the idea of digital immortality we lose what makes us human—that is, the experience of our multi-sensory engagement of being-in-the-world. This is the lived-event itself, the intensities of the dirt, noise, and sweat of life, not the rationalized pure signal of its digital record. As the philosopher Richard Rorty put it: 'If the body had been easier to understand, nobody would have thought we had a mind.'[18] Describing the pleasures of phonography, Evan Eisenberg cites how Odysseus 'leaves his immortal lover, knowing that time and ageing will make Penelope loveable in a way impossible for [the god] Calypso'.[19] A digital trace is truly undead, promising immortality not because it never dies but because it never lives. Eisenberg continues, 'the meaning that needs mortality, [feeds] off fading things'.[20] Through ideas of digital immortality corporations invite us to suffer the fate of all those who aspire to be gods; to crash and burn, like Icarus. We only have to be human to refuse this invitation.

17. P.D. Nunn and N.J. Reid, 'Aboriginal Memories of Inundation of the Australian Coast Dating from More than 7000 Years Ago', *Australian Geographer* 47:1 (2016), 11–47.

18. R. Rorty, *Philosophy and the Mirror of Nature* (New Jersey: Princeton University Press, 1979), 239.

19. E. Eisenberg, *The Recording Angel: The Experience of Music from Aristotle to Zappa* (New York: Penguin, 1987), 254.

20. Ibid.

DEATH BY EURO

Steve Goodman

'How do you kill someone who is already dead?' Nguyễn Văn Phong mused as he slumped over his desk. Morosely ensconced in a bland suite of a mid-range downtown Manhattan hotel, he spread and stacked the mini-skyline of 500 Euro bills in front of him, note by note, meticulously aligning each crisp, purple layer. How many notes would it take? He was unlikely to run out before he choked, as he had totally cashed out the IREX account.

It's New Year's Day 2002. Leaning left so that the full weight of his skull is supported on his elbow, Văn Phong is having one of his clearer moments of late, reveling in the fact that, momentarily at least, he again feels that he has internal organs. For around the last ten years he has been subject to horrifying waves of negation delirium. Episodically, this condition has led him to believe that he is merely a bloodless, disembodied simulation. Since 6 August 1991, Văn Phong has gradually started to tip into a profound depression that coincided with the escape of IREX[2] onto the embryonic internet. The daily administration of 50mg of Sertraline and 1mg of Risperidone had offered some respite, but the accumulation of these waves has landed him in his current impasse, leaving him existentially traumatized. The digital agency that had used him as a launch pad into global communication networks had no inbuilt safeguards, and Văn Phong strained under the weight of the catastrophic responsibility that was travelling back through time to meet him. Perhaps this burden had proved catalytic to the onset of his cybernetic strain of Cotard's syndrome—guilt at having unleashed a rogue artificial intelligence which, initially merely a tool, had come to psychologically dispose of its creator, hacking its own operating system. A binary genie with a set of objectives on some curve to infinity.

In the rush to forecast the magnitude of the financial earthquakes to be felt over the forthcoming decade, Văn Phong had ignored the threat posed by opening up such a portal to the future. From temporal paradoxes through to the super-optimized instrumentalization of humanoids, he had been left algorithmically unprepared. Now, four blocks from Wall Street, in advance of the drastic action he was about to take, he reflects on some of the Promethean encounters he has experienced since enrolling into this anomalous sonic research cell that had sent him burrowing down his current path to self-negation.

The first of these instances transported him forty years back, to his initiation into AUDINT and early conversations with his recruiter, Magdalena Parker, a few blocks away on the Lower East Side. He remembered Magdalena sharing whispered utterances, stories, and muddied firsthand accounts of golems that had become vivified through the simple utterance of words. She had also introduced him to a fascinatingly delusional, tin-foil-clad woman. Announcing herself as Rosalind Brodsky, with very little prompting she had unfurled a life-story of preposterous proportions. She spoke with certainty about the year of her death, 2058, and explained her presence there as the effect of her employ as time-traveller in the service of the Institute of Militronics and Advanced Time Interventionality (IMATI) based in South London, UK. Brodsky relayed how the Institute programmed virtual simulations of key historical events and, via the Hexen project, mapped, in the most abysmal detail, the flowchart of the military-occult complex. In 2039, she claimed, she had travelled to West Point US Military Academy, just north of Manhattan, to develop techniques of audio hypnosis and brain-modulating silent sound technologies. Many of the research findings from IMATI, Brodsky maintained, went on to inform British military mind control experiments between 2040 and 2045.

On another time loop, Brodsky, struck by vertigo from the aberrant temporal labyrinth in which she was now wandering, had sought relief from some of the major figures of twentieth-century psychoanalysis, undergoing therapy with Freud, Jung, Klein, Kristeva, and Lacan to decode her delusions. Clearly paranoid, she could often be overheard muttering about an artist, Suzanne Treister, whom she claimed was pulling her strings.

But the tale that had embedded itself most deeply in Văn Phong's memory was that of Brodsky's journey on another IMATI project that saw her travelling back in time to research the genealogy of the Golem. Infamously fabricated by a Rabbi Juddah Loew of seventeenth-century Prague to protect the local Jews during recurrent waves of pogrom, the Golem is a being from Jewish mysticism, made of inert matter, usually clay, and which is animated by a magical incantation.

More particularly, Brodsky had become obsessed by whether the proclivity to create artificial life had been genetically passed down from Rabbi Loew and was therefore a genetic rather than a social function. Her obsession had been ignited by the rumour that a number of key scientists in the field of AI, from John Von Neumann and Norbert Wiener to Marvin Minsky, Joel Moses, and Gerry Sussman, had been inspired by the myth of the Golem, all claiming to be direct descendants of the sixteenth-century Rabbi.

Almost twenty years later, in 1977, Văn Phong recalled, he had attended the fifth international joint conference on Artificial intelligence in Cambridge, Massachusetts. Trying to keep abreast of developments in computer science while working on his own IREX algorithm, he had met a group of Czech scientists (Bohuslav Kirchmann, Pavel Kopecky, and Zdenek Zdrahal) who had created a robot called the Goalem (Goal oriented electrical manipulator). Late at night, they had half-jokingly described to him the cybernetic kabbalism that lay behind their Goalem. In gematria, words are numerical codes. The idea had come to them that the creation of AI through codes, with silicon as base, paralleled the creation of the Golem through an incantation with clay as base. The utterance of the name of God functions as a cypher that unlocks, switches on the inhuman entity. Fabricating bodies from clay and inserting holy words into their heads to animate them was merely a precursor to constructing a robotic body and programming it to behave in a certain way. A glitch was merely the golem misinterpreting the words. Erasing the robot's memory was as simple as removing the words from the golem's head.

It was this idea of reverse engineering the Golem code that has haunted Văn Phong, offering a slender toehold on what seemed like his certain descent. Emotionally crippled by random, senseless bouts of Cotard's, he had fretted for years over the disengagement of his own Golem as it sped out of control. Every time he lost a sense of his own body, it concurrently felt like it was being uploaded by IREX[2]—and his body was a gradually dwindling resource. He remembers futilely making programming upgrades to IREX after possessed bouts of outsider trading. Using his third ear algorithms, he'd conferred with xenobuddhist consultant Jack Schwarz and psychobotanical high frequency trader Hillel Fischer Traumberg (twenty years into the future) about the dangers of fully automated trading systems gone AWOL. He'd even, in the mode of Rabbi Loew, tried to invert the programming so as to switch off IREX^2, but to no avail. It was too late. IREX^2 overcame any precautions and became self-governing, automating the weaponized waveform database of AUDINT and orchestrating a team of its own IRL meat puppets. IREX^2 was a superoptimiser, off the leash. Yet it was not IREX^2 as a malevolent singleton that would ultimately pose a threat, but the vortex generated by the twin system of IREX^2 and the THEARS, engaged in a competitive arms race.

The vaults of the Bank of Hell were opening. But instead of retiring his Golem, today Văn Phong would decommission himself. None of this would be his problem anymore if he submitted to the simulation, to which he had merely served as a host, in death. He was already undead anyway. The final straw had been the hectoring voice in his head over the last month projecting a twenty-year disintegration of the Euro, with cascading effects on global financial markets. Enough was enough. He threw on his coat and left the musty hotel, never to return. He knew exactly where he was going. Directly, it took him ten minutes. He found a spot outside Trump Tower on 40 Wall Street, shiftily opened a brown paper bag, and peeled off the first 500 Euro note. He no longer needed the body that for so long had been escaping from him. As he tentatively tackled the first note, his face souring, the noisy bustle muted. Even in the crowded street, the only sound was the deafening rhythm of his chewing and gulps. This would take some time.

ANCIENT TO THE FUTURE

Steven Shaviro

Music is an art of time, or of duration. It cannot be grasped synchronically, in a single moment. Indeed, music is only possible because we never live entirely in the moment. The Now, as we experience it psychologically, takes the form of what William James calls the *specious present*: a 'duration-block' within which 'we seem to feel the interval of time as a whole, with its two ends embedded in it'.[1] The present is 'specious' because it is never just itself. It always has a certain degree of thickness or spread. We grasp past and future, or before and after, together in a single, unbroken experience. Music gives us the intensity of a heightened present. But it also smears that present over a concrete temporal range. You can make a still of a movie image, but you cannot make a still of a sound.

Music addresses our experience of time. But our sense of time itself is subject to historical change. Over the course of the last three centuries, capitalism has contin-ually mechanized time, dividing it up ever more accurately, in ever smaller increments. Capitalism invented *metric time*, along with the technology for measuring it. This is time conceived as an empty and homogeneous linear succession of moments, with no qualitative differences between one instant and the next. In the early twentieth century, this process went even further, with the invention of the assembly line. Workers' actions and movements in the factory were regulated and micro-managed down to intervals of a second or less. The metric regularity of the capitalist working day is reflected in music with a regular metre.

Today, the capitalist appropriation of time has stretched far beyond the workplace. Labour time and leisure time can no longer be distinguished. We are always on call, 24/7. Production and consumption alike are organized on a *just-in-time* basis. There is

1.   W. James, *The Principles of Psychology* (New York: Dover, 1890), 609–10.

no off-switch. This leads to the shrinkage of horizons. The specious present is compressed into its narrowest possible limits. We are only able to think in the very short term. Ours is a society of attention deficit disorder (ADD). Everything happens quickly, and nothing lasts. 'Innovation' and 'creative destruction' are our watchwords. But ironically, this overarching condition of turmoil itself seems everlasting, unable to change. We have no future and no past, precisely because the past and the future have themselves been absorbed into the eternal present. The past is nothing more than a repertory of styles and poses, all of which can be sampled and remixed. The future, for its part, is pre-empted and colonized in advance; it is even given a present-time monetary price, through the trade in derivatives or 'futures contracts'.

Music has the power to push back against this regimentation of time. It can expand the reach of the specious present. Most obviously, it does this by preserving the past, by letting it resonate on every larger scales. This is why we cannot just consign the past to irrelevance. 'The past is never dead. It's not even past' (Faulkner). 'The tradition of all dead generations weighs like a nightmare on the brains of the living' (Marx). Far from being gone, the past is something that never goes away, because we are powerless to change it or repeal it. The past is immutable, irreparable. It's a wound that never heals without leaving a scar. Individuals may forget, but the world does not. Everything that happens leaves traces of its passing. Memory has an ontological consistency of its own; and music is the expression of this consistency. The past, as Gilles Deleuze puts it, is *virtual* rather than actual. Deleuze writes of two dimensions of time. One is Chronos, the time of the actual, of linear succession and of just-in-time production. But the other, deeper dimension of time is Aion, the time of the past preserved in its very pastness. Deleuze, quoting Proust, calls Aion 'a little time in its pure state', and a presence that is 'real without being actual'.[2]

Music can open us to the unknowable future, as well as to the immemorial past. But this requires a new articulation of time, one different from either Chronos (the time of the actual present) or Aion (the time of ontological memory or the virtual past). In music, both the past and the future are able, as Deleuze says, to 'elude [*esquiver*] the present'.[3] But they do so in radically different ways. There is no symmetry between the past and the future, because of the arrow of time: the fact that time is irreversible, and moves only in one direction. Where the past is closed, the future is open.

2.   G. Deleuze, *Difference and Repetition*, tr. P. Patton (New York: Columbia University Press, 1994), 122.
3.   G. Deleuze, *Logic of Sense*, tr. M. Lester and C. Stivale (London: Athlone, 1990), 1.

Where the past is an 'always-already', the future is a 'not-yet'. Where the past *subsists* as a trace, and the present *exists* as an actuality, the future *insists* as pure potentiality: as the incipience of a movement that may or may not ever take place.

Deleuze and Guattari, following Pierre Boulez, identify the music of Aion with 'the "nonpulsed time" of a floating music, both floating and machinic, which has nothing but speeds or differences in dynamic'.[4] But what is the music of the third sort of time, of incipience and open potentiality? Rather than being floating and pulseless, this music would have to be pulsed even to excess: a music of polyrhythms and syncopation. And indeed we have this music: it's the music of Afrofuturism, from jazz to funk to techno to hip hop. If the music of Aion is grounded in Bergson's notion of duration as melody, then Afrofuturist music is better understood in terms of Gaston Bachelard's revision of Bergson, which instead proposes multiple durations as rhythms. Afrofuturist music doesn't forget the virtual past—which it usually envisions in the form of the Egyptian pyramids. But it links this past directly to an equally virtual future: the time of space-ships and emancipated robots. From Sun Ra to George Clinton, and from Miles Davis's early-1970s electronic band to current Afro-divas such as Janelle Monae and Dawn Richard, Afrofuturist music disseminates new images of time: a time that is not only 'out of joint' (as Deleuze says of Aion), but also plural and unclosable.

---

4.   G. Deleuze and F. Guattari, *A Thousand Plateaus*, tr. B. Massumi (Minneapolis and London: Minnesota University Press, 1987), 262.

RAPPARITIONS

Toby Heys

The Holo Accords of 2056 map out an alternative constitution for discord management, a whole new way of engaging in conflict that reduces the massive costs involved in such endeavours and unconditionally removes flesh from the messy equations of political turbulence. From this point on all military operations will be conducted via holographic forces. Detachments, units, and divisions of encoded light fields, tactically mobilised for transparent effect. Gone are the days of collateral resource damage or civilian casualties, along with their subsequent cover-ups that reek like insipidly cheap perfume in the toilet of public opinion.

This is good for business, however you look at it, especially for newly emerging Holography companies. But this is not, in essence, a modern industry, not by any stretch of the imagination: Giambattista della Porta having first conceived of its inception by describing an illusion—'How we may see in a Chamber things that are not'—in his 1558 work of popular science *Magiae Naturalis* (*Natural Magic*).[1] Nearly three hundred years have passed by the time John Pepper and Henry Dircks manifest his idea in (virtual) reality by making a ghost appear onstage in Charles Dickens's theatrical rendition of 'The Haunted Man' in 1860.[2]

Two centuries on, the notion of conducting territorial, political, and natural resource struggle via holographic armies is a predictable extension and militarisation of the populist form of entertainment that projected itself into mass public consciousness in 2007: holographic concerts from musicians who had died and, more arrestingly, from those yet to be born. Fitting obliquely into the latter category is one Hatsune Miku,

---

1.   G. della Porta, *Magiae naturalis, sive, De miraculis rerum naturalivm libri IIII* (Naples: Matthias Cancer, 1558).

2.   For further reading on the evolution of Pepper's Ghost through to its holographic manifestation, see J. Steinmeyer, *The Science Behind the Ghost!* (Burbank, CA: Hahne, 1999).

a prophetic pop princess channelled by Sapporo-based Crypton Future Media.[3] With her vamped up Kabukichō stylee and cerulean pigtails she could not be more aptly monikered: her name translates as 'first sound of the future'. She is the first truly digital 3D crush for a slew of Japanese fans, and her presence works the salivary glands of technologists, teenagers, and posthumanists alike.

Following closely, as ever, US companies respond. In the first event of its kind, a dead rock star is brought back to life with voodoo fidelity, as the exhumed holographic corpse of Elvis Presley performs a duet version of his 1968 hit 'If I Can Dream' with Celine Dion on the TV show *American Idol*.[4] Through this endeavour, North American holotech industries spell out their rationale for mapping out the emerging era of the wraith as, pixel by pixel, they disinter the dead.

The year 2012 is ground zero for the popularisation of holographic projections or 'original virtual performances' as they are sometimes referred to in this era.[5] The Digital Domain Media Group revivify the rapper 2Pac in order for him to play live from the grave alongside Snoop Dogg (who claimed the encounter was 'spiritual') and Dr. Dre at the Coachella festival in California. In his own inimitable way 2Pac intones to the audience: 'To lead the wild into the ways of the man. Follow me; eat my flesh, flesh and my flesh'.[6] A zombie-call for future bloods to become immortalised by digital divinities.

Initially there is some unease about the sanctity of the posthumous performance of hits such as 'Hail Mary',[7] but this is blacked out by a public desire to bring young African Americans back to life after they have passed away at an unseasonably young age. The 2Pac production is quickly followed by zombie cameos from Ol' Dirty Bastard[8] as he joins Wu Tang Clan on stage to perform 'Shame on a Nigga'[9] and 'Shimmy Shimmy Ya'[10] at the Rock the Bells Festival, and from Eazy E, who appears with Bone Thugs-N-Harmony in 2013.[11] Then to cap it all, the king of the dead, Michael Jackson, is

3. See <http://www.crypton.co.jp/miku_eng>.
4. Elvis Presley, 'If I Can Dream' (RCA, 1968).
5. L. Zoladz, 'Ghost Riding', *Pitchfork*, 21 November 2013, <http://pitchfork.com/features/ordinary-machines/9265-ghost-riding/>.
6. Tupac Shakur, 'Hail Mary' (Death Row Records, 1997).
7. Ibid.
8. The 'appearance' of ODB took six months of labour for six minutes of airtime, according to his digital creator Chris Romero. See Zoladz, 'Ghost Riding'.
9. *Shame on a Nigga* was originally released as track 2 on side 1 of the vinyl *Enter the Wu-Tang (36 Chambers)* (Loud, 1993).
10. 'Shimmy Shimmy Ya' is the second single released from Ol' Dirty Bastard's first solo album in 1995, *Return to the 36 Chambers: The Dirty Version* (Elektra, 1995).
11. Together they performed the tracks 'Straight Outta Compton', 'Boyz-n-the-Hood', and 'Foe tha Love of $'.

brought back to life to perform in Cirque de Soleil's extravaganza 'One' at the Mandalay Bay Hotel in Las Vegas: now the man on the other side of the mirror, returning as the transposed picture of Dorian Gray.

The emergence of Holotech culture and the Lazarian industry it spawns in the USA are the final parts of the fiscal equation that multiplies young African Americans (especially those that are difficult to manage when still alive) with the morgue. The future figures of the body (and the income that will be accrued) rotoscope an amortized economy in which 'not only the labor but the laborer himself has been rendered immaterial, conjured up, and put to work. Outsourcing here takes on the character of "outsorcery", a conjuring of the dead to do work once the sole province of the living'.[12]

There is more to be made when the redivivi are birthed in light so that they can once again render material (wealth) through sound. In the 2020s the holo industry expands exponentially, and a rapparition index that hierarchizes hip-hop stars according to their estates and earning rates after death is established. Only those towards the top of the list will be holographically resurrected. Ghost money[13] adorned with Daedalian patterns and the revenant outlines of scrubbed throw-ups is supplied in printable format with downloads of albums and singles. When the currency is burnt by the purchaser, phantom royalties are paid, improving the rapper's social standing in the afterlife. On the dead presidents' face of it, the dollarization of Hell money has been jacked and diffracted; time to scratch off the worn heads of the Benjamin Franklins.

Each printable piece of ghost money is algorithmically illustrated according to the musical note count of the track or album it relates to. The three dominant musical notes in a track or album are translated into visual patterns by a CymaScope,[14] the normal mode of the music being broken down into a trio of granulated snapshots that resemble symmetrical spider webs by the way of orange sunshine. These patterns also contain encrypted data relating to the overall musical note counts that exist in the audio. Given that the drill-down density of each track's acoustic profile, this means that no two Audio Data Maps (ADMs) can be identical; their note and frequency-based information symbiotically form a highly complex system of encoding. The cymatic patterns

12.   J. Freeman, 'Tupac's "Holographic Resurrection": Corporate Takeover or Rage against the Machinic?', *Ctheory* (2016), <http://ctheory.net/tupacs-holographic-resurrection-corporate-takeover-or-rage-against-the-machinic/>.

13.   Also referred to as Joss Paper or Hell Money, Ghost Money comes in sheet paper format (currency) or as crafted objects such as smart phones. Used predominantly in China and Vietnam, it is burnt as an offering to the spirit of a deceased friend or relative, in order to improve their social standing in the afterlife.

14.   In the words of the Cymascope web site, 'A Cymascope is the first scientific instrument that can give a visual image of sound and vibration in ways previously hidden from view', < http://www.cymascope.com/>.

produced by this process are subsequently laid over a ghosted bomb[15] that is relevant to the music, and the overall design finally printed onto joss paper bills.

Each digital track is treated as an oscillating system that has its own waveformed logic, all parts moving sinusoidally in reaction to a frequency and fixed phase relation. At odds with this system of waveformed teleology is the value of the hell notes. While the price of the music is set, the value of the corresponding ghost money is not fixed, at least not until the consumer decides how much of their cryptocurrency they wish to part with. The final sequence of the purchase involves the reckoning of the phantom royalty, which must be made above the fluctuating minimum base rate, itself dependent on how the relative artist is currently ranked within the Rapparition Market Index.

The smoke created by the burning of the ghost money is captured by a remote sensory system called a Polsen—a satellite and aircraft technology that was once used to measure respirable suspended particulate matter. Here it is retooled and miniaturised for use in handheld devices on the ground. These small pieces of apparatus can be jacked so that they become ultra-sensitive to temperature as well as to sonic and optical information. As a result they are capable of detecting nanoscopic deviations in air quality and the patterns produced by gaseous emissions. When a paper offering is burnt, the Polsen detects the patterns of the gases and particulates so that, in effect, the ghost money emits cymatic smoke rings, which are captured and digitised. This data subsequently feeds back into the Rapparition Index, so that the phantom royalties can be tracked. Modal vibrational phenomena dictates holo market flow in this way, the liquid agency of the rapparition being based, in part, upon their smoked assets.

As well as the rapparitions' standing in the afterlife being improved (depending on how much of their ghost money is burned), their futures market is also adjusted and down- or upgraded accordingly. The higher their stock climbs, the more investment goes to fund the 'reality engine' of their holographic form, which, as a result, will become increasingly stable, lifelike, and, of course, ubiquitous. The aim is to become a black chip investment, the highest economic honour that can be bestowed upon a dead rapper. In financial circles, the term 'catching a body' is reverse-engineered and given a holo makeover, killin' it in the afterlife being the first step to gaining pounds in the present. The spectranomics of ghost money, then, equate to a temporal spread spectrum. Somewhere between quietus and revivification, the rapparitions are finally getting paid in full.

15.   The outline of a piece of complex graffiti.

XFD

PASSAGE, TRANSFER, LIMBO

THE SOLITARY PRACTICE OF
THE VANISHING CONCERT PIANIST

Tim Hecker

In 1964, Glenn Gould, one of the eminent concert pianists of his era, shocked many by retiring from live musical performance at the young age of 32. Gould, a noted hypochondriac who would not shake hands with concertgoers, continued the evaporation of his physical body by no longer peddling his wares as a real-time instrumental virtuoso. He gravitated toward the private recording studio—a mediated space of musical expression that combined the utopian optimism of 1960s networked communications with a musical life of relative solitude. He was a pioneer of an emerging bedroom studio culture that promised autonomy but which also often yielded a crippling perfectionism. For Gould, it was a move that ultimately aided his retreat from human contact. This early paradigm of musical autonomy is also a gateway to understanding both the creative liberations and the forms of domination that emerged during this period of increasingly isolated yet also interconnected musical practice.

Gould's deep ambivalences towards public exposure rendered the studio the only possible site of artistic expression and personal salvation. He viewed this spatiality as a sort of embryonic insulation from the world, a laboratory of the late-night, which, while rewarding experimentation, sheltered him from the external pressures that stifled creative development: '[I]t's, quite literally, an environment where time turns in upon itself, where, as in a cloister, one is able to withstand the frantic pursuit of the transient, of the moment-to-moment, day-by-day succession of events.'[1] The availability of the studio allowed him to continue his musical career by other means. It presented a radical break as a physical liberation from performance, but also a continuity in the sense of a performance through the studio itself. During the post-performance years

1.   G. Gould, 'What the Recording Process Means to Me', transcript from CBS Masterworks, 26 July 1982, National Archives Canada (NAC), Glenn Gould Fonds, MUS 109 16:76.

his aversion to the physical public grew to the point where he would only pick up the awards offered to him if there was no public ceremony. According to Gould historian Kevin Bazzana, he would only interact with the public through the prostheses of electronic media. Solitude, the overarching paradigm of his beloved creative listener, was also the monastic inner life of his studio practice.

Glenn Gould's studio isolation can be seen as a technology of the self, as an instrument that provided the conditions for self-constitution and personal salvation. For Michel Foucault, these technologies are actions that permit individuals to effect their own means in order to attain 'a certain state of happiness, wisdom, perfection or immortality'.[2] He traced the notion of retreat back to the Stoics' *anachoresis*. A retreat into the country became a spiritual retreat into oneself, a daily ritual, not so as to introspect or get in touch with one's inner feelings, but rather so as to reinforce and remember rules of action. So these positive technologies are also dialectically coupled with a mechanism of ascetic domination. For Gould, the inscription of mechanisms of domination such as the increasing divorce from his own body, became more apparent as he executed a near total disappearance from public life. The studio was eventually the only possible mediator between performance and audience. In a letter to an apparent stalker, Miss Adele Knight from Watertown, New York, he expressed shock that she showed up uninvited at his country house in Canada. He asked to be free of bother from the outside world:

I am not about to change, or further justify my preference for a life of solitude [...] I do believe that we may all reasonably expect a life free of intrusions from outside, and I am, therefore, going to ask you to reconsider very calmly the nature of your actions these last several years and to refrain all together from your attempts to visit or contact me.[3]

Gould's striving for the 'cloak of anonymity' that the recording studio provided was realized in an increasingly profound sense as his disengagement from the world continued. In 1976 he took up part-time residence at the Inn on the Park, a hotel deep in the suburbs of Toronto. Up until the time of his death Gould used the hotel as an editing

---

2.   M. Foucault et al., *Technologies of the Self: A Seminar with Michel Foucault* (Amherst: University of Massachusetts Press, 1988), 18.

3.   Letter dated 19 September 1966. National Archives of Canada, Glenn Gould Fonds, MUS 109 31:30:31.

suite and occasional living space in order to work, uninterrupted, on projects of various sorts—as a studio, a living space, and almost a crypt.

The history of literature and the arts are replete with examples of reclusive artists. Most have been stories laced with nostalgia, rejection, and withered dreams. Very few proactively sought a retreat with the belief that a new technology would afford greater expressive capabilities, better distribution methods, and new aesthetic possibilities all the while offering a simultaneous life of solitude. This was a route to the 'clinical ecstasy' of performance by other means. But the network mediated autonomous music studio that emerged after the 1960s should be seen as a technology of the Undead—a place that gave life and took life, all the while whispering the sweet promises of immortality nurtured by recorded media.

GHOST IN THE MACHINE: HIKIKOMORI AND DIGITAL DUALISM

Lisa Blanning

For the past quarter century or so, some people in Japan, mostly young men, have been confining themselves to their rooms and withdrawing from society. Avoiding work and social situations of any kind while depending upon their families to support them, extreme cases have seen this tendency last for decades. This small, bedroom-sized space is the world of *hikikomori*, and while the term was coined and the trend first identified in Japan, its emergence probably facilitated by the nation's rigid culture of high expectations and strong family bonds, it's a phenomenon with ever-increasing international relevance.

Hikikomori live like ghosts, bound to one place as mere shadows of their previous or potential selves. They exist, but refuse to participate in living as we know it, instead inhabiting an in-between state—wilfully giving up their agency but remaining undead. And while there are many reasons someone with no other mental disorder—one of the hikikomori criteria—may detach from their life, it's easy to read it as fear or rejection of society, other people, or indeed anything that could be viewed as 'normal' or 'necessary,' such as jobs or relationships. And yet a voyeuristic dip into a hikikomori messageboard (in this case, the now-defunct English-language *Hikikichan*)[1] reveals all of the usual desires for human socializing. In one thread, an anonymous poster confesses his (at least we assume it's a male) joy at meeting a girl online and the comfort he takes from their conversations and her acceptance of him. But as months pass without word from his 'angel', his confusion turns to grief. Responders to the original post, all of whom are also anonymous, share advice and their own experiences. 'I talk to nobody in real life', starts the most recent post, concluding, 'talking to people [online] is basically the only

---

1. <https://hikkichan.com>.

"happiness" I can get'. So, if hikikomori are not prepared to participate IRL, perhaps they do their actual living virtually.

In *Welcome to the N.H.K.*, a 2002 novel by Tatsuhiko Takimoto which has been adapted into both a manga and an anime series, the main character Tatsuhiro Satō is a hikikomori with a tendency towards conspiracy theories who believes that the N.H.K.—the actual initials and moniker of real-life Japanese public broadcaster *Nippon Hōsō Kyōkai*—is really the shadowy *Nihon Hikikomori Kyōkai* agency. Satō's N.H.K. is no benevolent national broadcaster. Instead, it's a corporation with evil plans aiming to turn people into *otaku* (obsessive geeks/nerds) and hikikomori through anime, music, and other media. This mirrors the current popular belief that the internet is causing the hikikormorization of a generation of web addicts—in this case, Western millennials—making them unwilling or unable to socialize in person, instead preferring online interaction. (While hikikomori as a phenomenon started before widespread use of the internet, it's hard to dispute the theory that the latter enables the former. In Japan alone, the government's 2010 estimate was that 700,000 were people living as hikikomori, with 1.55 million more on the verge of becoming hikikomori.) 'Millennials don't have sex!', the newspapers scream.[2]

So the internet is a transmogrifier, with the power to change normal people into otaku and hikikomori. But simultaneously, it's a solution to the hikikomori problem, offering a lifeline of hope, human contact, and shared experience, albeit mediated, at a physical distance, through a screen, at a controllable rate of exchange.

While representations of life in this digital age are all around us and seem to be growing exponentially in number, perhaps nothing addresses both the human condition on the internet and Satō's hikikomori conspiracy as well as DVA [Hi:Emotions]'s recent album *NOTU_URONLINEU*. British producer DVA not only references the digital dualism created by existing both IRL (in real life) and via an online persona in the title of the record, but also introduces a familiar fiction into its concept. Hi:Emotions or H:E is both part of his artist name and the name of a mega-corporation that, as the press release for the album states, 'is slowly taking control of everything, and plan to eventually make all people live life under one brand in virtual reality'. Clearly, H:E is an extension of the N.H.K. updated for the virtual reality era, and DVA their musical agent in the conspiracy.

2.   D. Buchanan, 'Millenials don't have sex? Of course we don't', *The Guardian*, 3 August 2016, <https://www.theguardian.com/commentisfree/2016/aug/03/millennials-dont-have-sex>.

On one hand, we can view *NOTU_URONLINEU* as the sonic manifestation of the hikikimori life—or rather, its undead state of in-betweenness. Listening to music in the dark was one of artist DVA's inspirations; he then created the record entirely in the dark, which we can take to mean that it should be listened to in the dark—the kind of activity hikikomori and otaku enjoy. And while DVA's background, back catalogue, and peer group are at home in the club environment, *NOTU_URONLINEU* is the sound of nervous energy synaptically firing in cyberspace. Though the club feels present—we can't shake off the memory of our bodies—mostly it's a reclusive exploration of interiority. Voices do appear, but they're mostly disembodied computer sequences, film clips ('*I don't love you anymore*'), or corporate advertising. When, in the track 'ALMOSTU', we hear the soulful, living mellifluousness of singers Rae Rae and Roses Gabor, their voices cut right through—angels piercing our solitude who make one lone visit and depart as quickly as they arrived.

If there's something unknowable in *NOTU_URONLINEU*, Puri Puri Puririn—the magical girl anime of which Satō's otaku neighbor Kaoru Yamazaki is a fan—is familiar and comforting to the max. Yamazaki listens to her theme song—a bouncy, fun J-pop confection—incessantly. As it oozes between their shared wall, Satō does, too. At first he hates it, but it's not long before it's both an involuntary earworm and his ringtone. Perhaps the sound of hikikomori is not so much the music of digital solitude, but instead a manufactured cheerfulness that you have no choice but to listen to. The music does not play for you, it's an echo of somebody else's life washing over the echo of your own life. Undead, you have no agency. No wonder you withdraw online.

'[DVA's] album hints at themes of online alienation, confusion, control and domination', the press release continues. '[It's] peppered with hints of faux-therapeutic advertising and psychotic jingles—a reflection of the stress of online life'. But track titles such as 'MEMORIESOFOFFLINEACTIVITY' suggest that online life is all that's left to us. DVA, at least, is completely transparent in revealing that even as we disengage from society and other people, retreating online, we retreat into a corporate sanctuary. Conceptual poet and university professor Kenneth Goldsmith, teacher of a course called 'Wasting Time on the Internet' at an Ivy League college in the United States, would agree. In an interview with UK newspaper *The Independent*, he explains:

> I have far-left friends who complain about corporate culture, and they do it on Facebook. They're not seeing that the platform is owned by something else. I think we want to

make technology transparent [but] it's highly intertwined with technological corporate advertising money that problematise that transparency. When it suddenly becomes visible, you go, 'Ugh, how did I not see it?' But at this point, you can't walk away. You have to be really privileged to walk away from digital culture. Social contacts, dating, jobs, everything comes through that.

If we are not all hikikomori now, interacting with others virtually through the mediation of the internet and in thrall to a corporation, perhaps the best we can hope for is a Pokémon-Go-style augmented reality. Goldsmith again: 'That's why Pokémon Go is so marvellous, it puts the body back into the landscape.' Writing for *The Society Pages* cyborgology blog, sociologist and social media theorist Nathan Jurgenson refutes the idea of digital dualism, in favour of augmented reality:

> Our Facebook profiles reflect who we know and what we do offline, and our offline lives are impacted by what happens on Facebook (e.g., how we might change our behaviors in order to create a more ideal documentation). Most importantly, research demonstrates what social media users already know: we are not trading one reality for another at all, but, instead, using sites like Facebook and others actually increase offline interaction.

Online life sapping offline sociability, forcing people to seek connection online, which in turn drives offline interaction—less of a paradox and more of a vicious cycle, perhaps. The transformational power of the internet is yet to be fully gauged, but clearly the N.H.K./H:E has found the most potent tool imaginable. We wrestle with ourselves not to succumb to their temptations: media wealth far beyond any otaku's wildest fantasies, digital worlds more navigable and accepting than our own. Augmented reality—incorporating the virtual alongside life in our own bodies in physical space outside of the confines of our homes, on a plane other than that of cyberspace—offers us a hikikomori-friendly alternative.

A BRIEF DEFENCE
OF NEW AGE AUDIO

Erik Davis

There may be no contemporary musical genre that is easier to accuse of false consciousness than New Age music. The anodyne synth washes, the timbres chiming like
floorwax, the easy pickpocketing of Native American drums and Japanese *shakuhachi*,
the avowed intention to calm, soothe, and spiritualize—all seem to materialize a music
of accommodation, passivity, and hazy escape, a music in which, as Marx wrote, all
that is solid melts into air. More than any genre besides the mall music they sometimes
resemble, New Age recordings seem in lockstep with both the self-serving affect
management of late capitalism and the neoliberal gospel of consumer self-realization.
A spontaneous revulsion seems to arise in many an engaged and critical listener of this
stuff, an impulse to reject not only the sounds themselves but their seeming social and
psychological function as well.

But there is something radical in this music, a radicalism that not only saturates
(select) works of New Age composition but informs the modes of listening and thinking they invite. Consider some of this populist genre's more avant-garde imperatives:
the dissipation or destruction of melody; the immense slowing and interruption of
temporal development; the construction of environmental *topoi* and other posthuman
ecologies; the multiplication of harmonic layers towards resonant chaos. Moreover,
such experimentalism is directly tied to both a DIY logic of technologized 'home-brew'
composition and, more importantly, an audacious bid to directly and immediately
deconstruct, reformat, and transform the subjectivity of listeners. As one of the first
explicitly psychotronic popular musics, New Age was composed and deployed with
the subliminal affordances of psychoacoustics explicitly in mind, at least rhetorically,
and it held as its ultimate goal the ecstatic dissolution of the bounded personal subject.

As is often the case, the most visionary possibilities of the music appeared during its salad days. Important progenitors arose in the mid-1970s, before the awareness of genre had crystallized into an alternative market and before it earned its name—always contested by some makers. Indeed, the recent revival of interest in the genre by collectors and curators in part reflects the recognition that early works by Iasos, Michael Sterns, Larajji, Joanna Brouk, Deuter, and Upper Astral remain fresh and inspiring engagements with sonic possibility, crafting new forms by melding already flowing currents of Krautrock, minimalism, psychedelia, jazz fusion, Asian traditions, and analog electronica. This 'golden age' lasted roughly until the mid-1980s, though glints of it still glow today, in the cracks and fissures between all manner of musics that spatialize, reverberate, and dissolve.

For all its rhetoric of oneness, much New Age music operates on at least two levels, one signifying and one asignifying. As an aesthetic object or audio representation, the New Age piece serves as an audible model for a possible internal environment, a kind of sonic model of a possible altered state within the listener, a state that would realise itself through the listener's merger with the audio. Take flautist Larkin's 1980 composition 'Emergence', which combines the sounds of the ocean (waves, humpback whales) with gentle aimless instruments, drawing the listener into a seaside trance. The heroic self of early Romanticisms has dissolved into an elemental fusion, even as the subtle presence of a slowly pulsing electronic tone rhetorically disguises (and subtly affirms) the recording's dependence on the technology that might otherwise be seen as intruding on the scene.

As a representation, this sort of fusion might be considered 'bad faith'—even though the gestures of naturalist intimacy should not be discounted in an era of claustrophobic ecological collapse. Hovering beneath this representational level, however, New Age takes a much more posthuman stance. Beneath its surface shimmer, New Age music declares and deploys itself on another level, a psychoacoustic or subliminal domain of vibrational control over hidden, unconscious and sometimes mystical systems of the human bodymind. In other words, composers intentionally and sometimes explicitly present their music as an apparatus for directly operating on the expanded psycho-bio-cosmic systems that are 'running' the programs of conscious awareness. Whether or not this self-presentation is merely rhetorical, New Age music should still be seen as part as a kind of utopian or idealist posthumanism, in which audio is optimistically deployed and consumed beyond the limitations of the personal listening subject.

This embrace of the vibratory unconscious was shaped by the aspirations towards human potential associated with the 1970s, when the evidence from experimental psychology and the transpersonal altered states crew alike suggested that the liminal zones of human consciousness were amenable to control and manipulation. Perhaps the bestselling independently pressed New Age record remains Stephen Halpern's often re-recorded 1975 'Spectrum Suite', which was fashioned with subliminal intentions, as explained on the back cover. For Halpern, the transformative—or at least 'altering'—effects of the music lay partly in its abandonment of melodic and harmonic drive. Elsewhere Halpern called it a 'non-anticipatory' music, or, more amusingly, as a performance of *scalus interruptus*. Halpern was also one of first musicians to study the effects of music with biofeedback and EEG, though he leapt well beyond instrumentation in linking specific pitches to colours and the esoteric physiology of human chakras.

Of course, both the scientific and the commercial aspirations of all this can look pretty sad in the light of contemporary affect management and the widespread incorporation of holistic practices into capitalist enterprises. How could the production of manipulating audio designed to soothe and relax not represent the sort of operational post-ideology required to keep producers and consumers tethered to a machine whose increasingly inhuman demands and pace are already pulverizing subjectivity?

But New Age audio, at least in its most experimental and intensive forms, is also a music of operational ecstasy that does not soothe the subject so much as teach it to probe its own inevitable dissolution—in death, in merger, but also in those ecological and cultural collectivities without which we are toast. In that light, and especially given its proactive embrace of technicity (electronic gear, binaural beats, subliminal controllers, etc.), New Age audio can also be understood as a kind of cosmic Accelerationism. The only solution to the posthuman crisis of subjectivity is through and beyond it—Burroughs's *here to go*—a deceptively gentle radicalism that dissolves the boundaries of subjectivity not through violence but through the subliminal audio body's capacity to rewire affectivity and awareness from within, and, more potently, from the sub- or superliminal *in between*.

PURGATORY

Nicola Masciandaro

> E'en then, when from its neck of marble torn,
> His head Oeagrian Hebrus bearing down
> Its central current rolled, 'Eurydice',
> The voice itself and death-cold tongue—alas!
> His poor Eurydice with fleeting breath
> Was calling still.
>
> — Virgil, *Georgics*, IV.523-6 (trans. Sewell)

The capacity of Orpheus's severed head to sound among the living from beyond the threshold of death indexes a dimension of voice and sound, or more properly an indetermination of the two, which may be defined as purgatorial. As the poet's words make clear, it is not the person here who speaks but voice itself [*vox ipsa*], less the sonic presence of a being than the more uncanny call of the very slipping away of its soul or life [*anima fugiente*]. This liminal sound, voiced at once through and without the tongue that articulates it, is purgatorial in several senses: (1) in connection to the Orphic desire to save a beloved from the death's underworld; (2) in its connotation of a self-purifying spiritual suffering at/of the limits of being; and (3) in its hyper-actual virtuality, the weird or haunting phenomenal immanence of something in the seeming absence of its own possibility. How does this paradigmatic 'third place' as Luther called it, a dimension characteristically neither provable nor deniable and persistent within modern culture in the undead forms of the medieval imagination, belong to what this volume's editors call 'the broader vibrational continuum of which perceptible sound is only a subset [...] a third dimension in which the real and the imagined [...] bleed into one another'?[1]

---

1. The obscurity of the fact of Purgatory is expressed as follows in the second Appendix to Aquinas's *Summa Theologica*: 'it is sufficiently clear that there is a Purgatory after this life [...] Wherefore those who deny Purgatory speak against the justice of God [...] Nothing is clearly stated in Scripture about the situation of Purgatory, nor is it possible to offer convincing arguments on this question' (Thomas Aquinas, *Summa Theologica*, tr. Fathers of the English Dominican Province [New York: Bezinger Brothers, 1947], 3022–3). Similarly, before rejecting the doctrine, Martin Luther wrote,

Mediaeval discourse on purgatory revolves around its ambivalence as both place and state.[2] Bonaventure writes, 'As for the state of purgation, this corresponds to an inde-terminate place [*locus indeterminatus*] in relation to us and in itself'.[3] Dante's Purgatory, an island orogeny in the southern hemisphere caused by Lucifer's fall to the center of Earth, finds itself at the summit of this ambivalence. Here the truth of purgatory and the truth of poetry converge in a reality as concrete as it is fabulous, a geography of the imagination in both senses. Purgatory is the ground of poetry's own resurrection: 'But here let dead poetry rise again' (*Purgatorio*, 1.7). Significantly, this resurrection is not only spiritual but specifically sonic: 'and here let Calliope arise somewhat, accom-panying [*seguitando*] my song with that sound [*suono*] of which the wretched Magpies so felt the blow [*colpo*] that they despaired of pardon' (*Purgatorio*, 1.9-12).[4]

Synthesizing the penitential and resurrective poles of Purgatory into something trembling within-beyond the threshold of representation, the central sonic image in Dante's *Purgatorio*, which figures the pilgrim's experience of the sound of its very gate, is paradoxically thunderous and harmonious, harsh and sweet:

'The existence of a purgatory I have never denied. I still hold that it exists, as I have written and admitted many times, though I have found no way of proving it incontrovertibly from Scripture or reason.' (*Luther's Works* [54 vols. Philadelphia: Fortress Press, 1957], vol. 32, 95). On purgatory in the modern world, see R.K. Fenn, *The Persistence of Purgatory* (Cambridge: Cambridge University Press, 1995); J.L. Walls, *Purgatory: The Logic of Total Transformation* (Oxford: Oxford University Press, 2011); and S. Greenblatt, *Hamlet in Purgatory* (Princeton, NJ: Princeton University Press, 2013). The place of purgatory in the history of Western music is correlatively double-sided. One the one hand, purgatory, as the theological ground of intercession for the dead, is a monumental potential of musical creativity. In the mediaeval period, 'Purgatory encourage[d] endowments supporting polyphony' and influenced the musical development of votive Masses (B. Haggh, 'The Meeting of Sacred Ritual and Secular Piety: Endowments for Music', in T. Knighton and D. Fallows (eds.), *Companion to Medieval and Renaissance Music* [Berkeley, CA: University of California Press, 1992], 64). Most prominently, the requiem mass 'realized a privileged status in music history [...] [exercising] a prominent influence upon subsequent musical styles, both sacred and profane', and '[t]hroughout the seventeenth century, musical settings of the requiem mass spread like wildfire as hundreds of new settings were composed' (R. Chase, *Dies Irae: A Guide to Requiem Music* [Landham, MD: Scarecrow Press, 2003], xv–xvii). On the other hand, the post-Reformation demise of the doctrine of purgatory was itself impetus for musical invention, for the development of alternatives to the requiem mass and new musical forms in the context of 'the Reformation [which] saw many of the sounds of death removed', most conspicuously the death knell (D. MacKinnon, '"The Ceremony of Tolling the Bell at the Time of Death": Bell-ringing and Mourning in England c. 1500–c.1700', in J. W. Davidson and S. Garrido (eds.), *Music and Mourning* [London: Routledge, 2016], 34). The doubleness of this reflexive relation between music and purgatory is reflected in Luther's famous reverence for music as 'next to theology' and recognition of its spiritual power: 'For we know that music is odious and unbearable to the demons [...] [music] alone produces what otherwise only what theology can do, namely, a calm and joyful disposition' (quoted in R.A. Leaver, 'Luther on Music', in T.J. Wengert (ed.), *The Pastoral Luther: Essays on Martin Luther's Practical Theology* [Grand Rapids, MI: Eerdmans, 2009], 271, 285).

2.   See J. Le Goff, *The Birth of Purgatory*, tr. A. Goldhammer (Chicago: University of Chicago Press, 1984).

3.   Bonaventure, *Commentaria in librum quartum Sententiarum*, Quaestio II, quoted in Le Goff, *The Birth of Purgatory*, 253–4.

4.   Dante Alighieri, *Purgatorio*, ed., tr. R.M. Durling (Oxford: Oxford University Press, 2003).

Then he [the angel] pushed open the door of the blessed gate, saying: 'Enter; but I warn you that whoever looks back must return outside.' And when the pins turned in the hinges of that sacred palace, pins made of strong, resonant metal, Tarpeia did not roar so nor seem so harsh [...] I turned attentive to the first thunderclap, and I seemed [*mi parea*] to hear voices, singing '*Te Deum Laudamus*,' blended with the sweet sound. The image [*tale imagine*] rendered in what I heard was exactly what one perceives when there is singing with an organ so that now one understands the words, now not [*or sì or no*]. (*Purgatorio*, 9.142-5)

As if echoing with both the roaring superessential voice of God to be heard at the end of time (Revelation 14:2) and the angelic harmony of the cosmic spheres whose motion is time itself, this sound, at once a sound passed through and the sound of that passage, is an opening audible between time and eternity. Yet as the common analogy emphasizes, this salvific opening is nothing abstract or otherworldly, but a palpable intensity located in the negative continuity of sonic seeming, in the intangible space between voice and hearing, the moving indistinguishability of words and music. Just as a soul may be saved by 'one little tear [*una lagrimetta*]' (*Purgatorio*, 5.107), so the essential sound of Purgatory is crucially a movement within the moment of experience. Furthermore, the poet aligns this momentariness with the instantaneous event of the image within the sphere of hearing, implying a continuity between Purgatory's gate and the senses known to William Blake: 'If the doors of perception were cleansed every thing would appear to man as it is: infinite.'[5]

Like a sonic intensification of the dialetheia of the image, which is always both true and false, both what is seen and what is not, the resonance of purgatorial opening is something heard and grasped in a movement that must, like Orpheus returning with Eurydice from the underworld, not look back—except, of course, through the special retrospective lens of poetry as a privileged labour of love which, inspired from above, is saved from its own oblivion. For only the poetic image, itself always both true and false, can enter and pass through the mirror of the imaginal realm, without breaking it, as it were. Only the musico-fictive third of sound and word can speak to and from the depths, touching what is otherwise invisible, like the self-doubling purgatorial voice of Poe's Valdemar: 'Yes;—no;—I have been sleeping—and now—now—I am dead.'[6]

---

5.   W. Blake, *Complete Poetry & Prose*, ed. D.V. Erdman (New York: Doubleday, 1988), 39.

6.   E.A. Poe, 'The Facts in the Case of M. Valdemar', in *The Complete Tales and Poems* (New York: Vintage, 1975), 101.

The abyss opened by the purgatorial resonance of the poetic word is the sound of the present itself as the perpetual intersection of the *nunc fluens*, the temporal now that passes and the *nunc stans*, the eternal now that stands. This intersection is not a point or instant, but a dilation: 'For the gate is narrow and the way is hard, that leads to life, and those who find it are few' (Matthew 7:14)—its narrowness and hardness being not only the steepness of ethical and spiritual becoming, the need for renunciation and surrender, but more immediately the continuous and ever-dilating nature of the gate itself which, reverberating with a spontaneous and unmasterably positive/negative sound, demands a proportionate spontaneity of doing and not-doing, a daring-still pace that can quickly-slowly keep time with the spontaneous rhythm of the eternal or trace the style of God.[7] Thus the deeper horror to which Poe's story gives voice is not the fear of whether there is or is not an eternity somewhere over the rainbow, but the terror of its being here now, actually present in the midst of time and too close for the comfort of all-too-temporal human identities. Correlatively, Orpheus's failure represents the loss of the present itself, its being shrunk to an instant, as Ovid's words make clear:

> At once [*protinus*] she slipped away [*relapsa est*]—
> and down. His arms stretched out [*intendens*] convulsively
> to clasp and to be clasped in turn, but there
> was nothing but the unresisting air.[8]

Here the lover loses his beloved all over again in a movement that beautifully embodies the very nature of the living present which his intention, in seeking to grasp it, negates. As Eurydice's slipping away or relapse is itself an instance of the immediately infinite momentum or continuously forward movement of time (*pro-tinus*, literally a reaching forward or onward), so does Orpheus's futile embrace figure the stretching of the present into the now. If only Orpheus had used his ears rather than his eyes. To pass through Purgatory, to cleanse—by passing through—the rusty doors of perception opening to infinity, involves hearing a sound that follows one ahead into forever, a music which, moving backwards and forward in time, is the reverberation of the present as something to be simultaneously lived and remembered, that is, experienced in

7. On the 'narrow gate' as the ever-present dilation of present found in the absence of worry, see N. Masciandaro, 'The Sweetness (of the Law)', in *Sufficient Unto the Day: Sermones Contra Solicitudinem* (London: Schism, 2014), 6–42.

8. Ovid, *Metamorphoses*, tr. A. Mandelbaum (New York: Harcourt, 1993), 327 (X.55–9).

the full-emptiness and empty-fullness of its ever-dilating nature. As Meher Baba says, 'Live more and more in the Present which is ever beautiful and stretches away beyond the limits of the past and the future',[9] and '[r]emember the present in the frame of the past and the future'.[10]

Unbounded by its theological doctrine, the sonicity of Purgatory stretches historically beyond mediaeval ghost stories and prayers for the dead, backwards into the speaking severed heads of antique and hagiographical legend and forwards into the spectral voices of modern surrealism and horror.[11] What is at stake throughout this domain is the interface between sound/voice and the outside of time, an interface unveiled in the deep *presence* of sound, in its being, through its very movement, something both within and beyond time and space. Similarly, Eleni Ikoniadou speaks of the 'concept of rhythm' as belonging 'to the middle, unleashing the relational potentialities of the notion of the gap and mocking the idea of distance as a void'.[12] The idea of Purgatory resounds with a future beyond temporality, beyond the division of life and death. As Chateaubriand observed, 'Purgatory surpasses heaven and hell in poetry, because it represents a future and the others do not.'[13] If this is a fire worth losing one's head over, it is because the promise of friendship or love—with anyone and/or Reality itself—requires it. As Blake testifies:

I have tried to make friends by corporeal gifts but have only
Made enemies: I never made friends but by spiritual gifts,
By severe contentions of friendship, and the burning fire of thought.[14]

9. M. Baba, *The Everything and the Nothing* (Beacon Hill, Australia: Meher House Publication, 1963), 37.

10. M. Baba, *Not We But One* (Balmain, Australia: Meher Baba Foundation, 1977), 52.

11. See J.-C. Schimitt, *Ghosts in the Middle Ages: The Living and the Dead in Medieval Society* (Chicago: University of Chicago Press, 1998); A. Simon Mittman, 'Answering the Call of the Severed Head', in L. Tracy and J. Massey (eds.), *Heads Will Roll: Decapitation in the Medieval and Early Modern Imagination* (Leiden: Brill, 2012), 311–27; R. Mills, 'Talking Heads', inC. Santing, B. Baert, and A. Traninger (eds.), *Disembodied Heads in Medieval and Early Modern Culture* (Leiden: Brill, 2013); I. van Elferen, *Gothic Music: The Sounds of the Uncanny* (Cardiff: University of Wales Press, 2012); and A.S. Weiss, 'Death's Murmur', chapter 2 of *Breathless: Sound Recording, Disembodiment, and the Transformation of Lyrical Nostalgia* (Middletown, CT: Wesleyan University Press, 2012).

12. E. Ikoniadou, *The Rhythmic Event: Art, Media, and the Sonic* (Cambridge, MA: MIT Press, 2014), 87–8.

13. Quoted as epigraph in Le Goff, *Birth of Purgatory*.

14. Blake, *Complete Poetry & Prose*, 251.

VR OF VOID

Alina Popa

— How are you, Madam?

— The no one of myself is not a Madam, call me Miss, please.

— I don't know your name, could you tell it to me?

— The no one of myself has no name: she wishes you didn't write.

— I would nevertheless like to know what your name is, or rather what your name was in the past.

— I understand what you mean. It was Catherine X.... We shouldn't talk about what has taken place. The no one of myself has lost her name, she gave it away upon entering Salpêtrière.

— How old are you?

— The no one of myself has no age.

> — Jules Cotard, *A Study of Neurological and Mental Disorders*

I'm ready to download eXistenZ,
by Antenna Research, into all of you.

> — David Cronenberg, *eXistenZ*

Cotard's syndrome was identified as a psychiatric condition in itself, distinct from the closely related hypochondria, by nineteenth-century Parisian neurologist Jules Cotard. Patients with Cotard's syndrome suffer from the 'delirium of negation', the active feeling that they do not exist, prompting in them a pathological (or perhaps sane) wish that the world should also disappear. Those affected by this rare neurological disease are stripped of what recent philosophy of mind would call 'the illusion of selfhood'. They are plagued by the conviction that they are being spatiotemporally deceived, since it seems to them that they are already long gone from this world:

One asks their name? They have no name. Their age? They have no age. Where they were born? They were not born. Who were their father and mother? They have neither father, nor mother, nor wife, nor children. Whether they have headaches, if their stomach hurts, if some part of their body hurts? They have no head, no stomach, some of them even have no body.... For some of them negation is universal, nothing exists anymore, they themselves are nothing.[1]

Even though the reality of their living body is palpable, they perceive themselves as alien; their body is an object, not a subject. They call themselves 'Zero' and 'X' and 'No one'. They do not possess a '*Körper*' but a '*Leib*', not an ensouled body but only a corpse; they are not 'I' but 'it'.

This syndrome has recently received special attention in consciousness studies and philosophy of mind: since there is a neurological disturbance that can plunge the human into the experience of her inexistence as a distinct self, then there must be neural correlates that generate the illusion of a global first-person perspective. The history of science in modernity has been a history of methodical suspicion toward the appearance of the world as perceived by a first-person subject. If knowledge is hostile to experience, it is this hostility that is being experienced by the zero-person subject in the here and now:

> Generally, the alienated are negators; the clearest demonstrations, the most reliable affirmations, the most affectionate gestures leave them incredulous and ironic. Reality has become strange and hostile to them.[2]

So 'the alienated ones' of Cotard's clinic are in fact among the few non-alienated ones, or are as little alienated as one can be, as far as the experience of oneself as self is concerned. In the mode of neurological disturbance that causes Cotard's syndrome, people are diagnosed with 'the madness of opposition'. But in the mode of normalcy, people may be terminally diagnosed with pronoun delusion.

The Melanesians of New Caledonia, as described by Lynn Margulis and Dorion Sagan in an article about the decentralization of the concept of self, are unaware that their body is an element they themselves possess. Likewise, Cotard's patients, divested of the phenomenal experience of their self, are deprived of the integrity of their body as a subjective body. Even though they are living outside of capitalism's assembly lines of

---

1. J. Cotard, *Études sur les maladies cérébrales et mentales* (Paris: J.B. Baillière et fils, 1891), 315 (all translations my own).
2. Ibid., 315.

canned selves and mechanized bodies, the Melanesians can be said to have become, paradoxically, the pioneers of the modern process of perspectival decentralization. Similarly, Cotard's patients are an embodied metaphor of the scientific dream of doing away with first-person perspectival bias. The constructing of an outside to the relatively homogeneous structures of experience, not only biologically entrenched but also culturally ossified (by Western global modernization) can be a matter of voluntary practice, the result of bodily dysfunction, or even an embodied theoretical or scientific exercise. Certain types of meditation practices, for instance, sharpen one's attention so much that one can overcome perceptual biases and experience the loss of one's own body, its dissipation into space. Some mystics become sensitive to subcognitive processes and, like Cotard's patients, end up speaking about themselves in the third person.

It is not only the clear-cut feeling of self that is at stake under biological and cultural Cotardian conditions; the realness-effect of the world itself may be altered with the onset of this affliction. Even if their 'I' has vanished, a virtual reality (VR) clings insistently to Cotard's patients' inexistence:

> It seems to the patient that the real world has completely vanished, has disappeared, or is dead, and that there remains only an imaginary world in the middle of which he is tormented to find himself.[3]

As Cotard himself reports, his patients are not naive realists. The ruses of their consciousness have been rendered opaque, revealing the holographic character of perceived reality. Their world is a species-bound theme park in which the techniques of make-believe have been left on show—a Brechtian theatre of consciousness. Cotard's syndrome patients are living in the first-person the virtual reality model of consciousness described in the third-person by philosopher of mind Thomas Metzinger. In his model, subjective experience is like a non-lucid dream, and the perceived world is a virtual reality. And since humans are bound to their immediate delusional perception, that of first-person experience, they can only self-induce Cotard's syndrome by theoretical thought or practical experience. There is only VR.

Scott Bakker's novel *Neuropath* presents us with two incarnations of the researcher. The first is Neil Cassidy, the thinker who not only thinks his thoughts, but also seeks to 'practice' them. The second is Thomas Bible, Neil's long-time friend, an academic

---

3.   Ibid.

whose life practices are disengaged from his ideas. They are both implicated in a theory of consciousness that revolves around 'the Argument'—an old discussion in which they had concluded that the brain is a spin-doctor of reality, a VR inducer, the constructor of a delusional first-person perspective. In *Neuropath*, Bible's disembodied theory becomes normativity and boredom, while Cassidy's embodied theory transforms into horror. Seeking to embody the Argument, Neil becomes involved in a monstrous quest to strip humankind of its default realism. He ultimately aims to rid his friend Thomas of his distance to the theory that he himself produced, and to which both of them remain, in very different ways, attached. At the end of the book, Neil torments Thomas by wiring his brain to a hallucination-based torture machine, shaking him out of the stable VR his academic and family life have become. In order to embody his theory, Neil tailors his neural circuits so that he can altogether abandon the fiction of being a self. He becomes 'no one' in order to exempt himself from empathy and guilt, a necessity if he is to open the world's eyes to the VR that it itself is. The neuro-horror of depersonalization tears down the fourth wall of the human world's real stage.

Thomas, the Cotardian academic, is tortured by the flesh that his argument has now become. Neil, the Cotardian artist (if we agree that the artist works with the abstract by manipulating materials, even including his own body), tortures flesh with the whip of thinking. The symbolic scaffold that keeps life separate from thinking creaks and squeaks when thinking laughs at life, howls and yowls when life grins back at thinking:

> Everything suddenly seemed at once fictional and impossible, like paint splashed across something monstrous. And quick, terrifyingly quick. Psychologists called such episodes 'derealization'. The irony was that they used the term to describe a kind of disorder, when it was about as accurate as any conscious experience could get.[4]

To deliver the Argument through his novel, Scott Bakker, the puppeteer of both torturer and artist, creates a fiction that flays our sense of self with the idea of the self's irreality. Upon entering this perspective-thriller, we get rid of our selves, but only by experiencing the book itself as a VR, one that leaves us no self:

> 'Depersonalization,' she repeated.

---

4.   R. Scott Bakker, *Neuropath* (Toronto: Penguin, 2009), Epub.

'You say that like it's a disease, but it's not, is it? It's more like some kind of…revelation or something.'[5]

The triptych of conscious experience as void, that of Neil, Thomas, and Bakker, presents one with the same paradox—isn't this experience, supposedly closer to that of the vigilant realists, just another Disneyland, eXistenZ, or virtual reality—a VR of the void?

The third-person viewpoint of a first-person perspective terrorizes experience. The first-person embodiment of a third person viewpoint terrorizes the concept. The experiment of entering a perspective that lacks perspectivality, that of the Cotard's syndrome patient, changes the third-person of science into a first-person of experience, and the first-person of experience into the universal zero-person. The lack of perspectivality goes deep, and all we can ever do is to change the depth of this void. For without 'I' there can be no 'it', and without 'it' there is no 'I'. Investigating Cotard's syndrome from a third-person perspective such as that of the present essay is as impossible as giving an accurate account of subjective experience from a scientist's perspective. To think that *the self is not* is, for a self, whether deluded or not, to enter a theme park or an inner VR of the void. Inhabiting the world as a delusion by deluding oneself into experiencing the void, throwing the self down the drain by flooding the mind with the illusion of the drain, is as close as one can get to the intuition of one's own inner ruses. The question posed by post-Kantian continental philosophy—that of access to the real—is implicitly also the question of unmediated access to one's inner ruses. But if all ruses were to be washed away, there would be no toying with knowledge anymore. So the anomaly is here to stay, and by investigating it we can only ever proliferate it. In proliferating the VR of the void, I am not emptying out the ruse, but filling it with void, while drawing close to the experience of realness. Overfilling the ruse with void may bleed some real through its grinning rifts. To access the ruse is to access the void is to access the real is to access the void.

Rumi was right to ask himself:

Who is the one in my ear
Listening to my voice,
Who is the one in my mouth
Saying my words?

5.   Ibid.

DELUSIONS OF THE LIVING DEAD

Toby Heys

It is 1949, and AUDINT's Walter Slepian, Bill Arnett, and Hypolite Morton are discussing how those with extreme psychological disorders might react to the tests carried out on their engineer Eduard Schüller, which have resulted in him conversing with the dead.[1] Meetings with Theodore Reik, who is researching for his book *The Secret Self*,[2] have been rich and varied. One line of enquiry involves the work of French neurologist Jules Cotard into a condition that leaves those afflicted believing they have no blood and that their body is without organs. Ultimately, it causes them to think that they are dead.[3]

When Reik tells of hushed rumours alluding to a notebook containing instructions on how to induce Cotard's syndrome, AUDINT are captivated. They speculate on how their Two-Ring table might be deployed upon those who believe they are already deceased. Would this alter the dynamics of communication with the otherworld voices? Could they transform carriers into necromancing drones by playing hooks from regular vinyl records?

Having spent months unearthing stories that corroborate a rumour locating the notebook in Paris, it is decided that Slepian, nominated for his patchy knowledge of the French language garnered from reading Proust's *À la recherche du temps perdu*,[4] will return to France for the first time since the Ghost Army departed after the Second World War. On board a TWA Lockheed Constellation he sits down, heaves a trepidatious sigh, and prepares for the twenty-hour trip from New York's Idlewild airport.

---

1.   Immediately after the Second World War, ex-AEG employee Eduard Schüller was sequestered by the AUDINT research cell and become a test subject for their Third Ear experiments.

2.   T. Reik, *The Secret Self: Psychoanalytic Experiences in Life and Literature* (New York: Farrar, Straus and Young, 1952).

3.   Subjects with Cotard Delusion, also referred to as Walking Corpse Syndrome, believe that they are already dead, have no blood, or have lost internal organs.

4.   M. Proust, *In Search of Lost Time*, ed., tr. C. Prendergast (London: Penguin Books, 6 vols., 2002).

His journey's reading consists of texts pertaining to Cotard, but it is a single piece of paper concerning a buyer of esoteric medical documents that absorbs him.

The information on the sheet concerning Isabelle Chimay is scant, consisting of a home address and the name of the 'Melac bar à vins' in the 11th arrondissement that she is known to frequent on weekends. That his trip is solely based upon staging a feigned serendipitous meeting with her at this haunt seems wildly optimistic, but it is all he has. It is early Thursday evening when he touches down on a cheaply perfumed bed in the illustriously shabby Alba Opera Hotel in the centre of Paris, and Slepian is bone-weary from the din of sleep-repelling propellers. Still exhausted, he spends Friday recovering, mulling over his plot to hornswoggle an unsuspecting collector of texts.

Whilst his knowledge of Cotard's work is his first weapon of seduction, his second is a vial of Amobarbital, otherwise known as truth serum, a drug used by the US military to treat shellshock so that soldiers could return to the front line. On Saturday night Slepian's front line is the Melac doorway, which he crosses at 6.35pm with a distorted gait owing to the bottle of wine he has already consumed. Inside, he sits close to the door in order to hear the verbal exchanges between patrons and staff, for he has no idea what Miss Chimay looks like.

Three glasses of a 1945 Pomerol Bordeaux in and Slepian, from behind a glazed patina of nerves, observes the entrance of a serenely upright lady. With a staccato sophistication fleshed out by rapid steps, her presence demands action. The maître d' snaps to attention, as do Slepian's senses upon hearing the greeting to Madame Chimay. Hands gesture towards a table that has obviously been kept for her. Slepian's plan was to wait for an opportune moment, but alcohol has sequestered his guile, so in broken French he brazenly introduces himself. With a confident quizzical smile Miss Chimay invites him to sit, and so the ersatz encounter begins.

They talk politics, music, and finally perceptual disorders—the phantom topic haunting Slepian's every word. Isabelle Chimay is forthright and passionate when revealing her penchant for rare medical documents. She states that her ability to speak English is the result of long hours spent translating letters from the 1860s. Ruminating on Phantom Leg Syndrome, the dispatches formed a small part of the Civil War correspondence between American poet Walt Whitman whilst he dressed wounds at Union hospitals and Silas Weir Mitchell[5] a 'doctor of nerves' from Philadelphia.

5.   Silas Weir Mitchell (1829–1914) was a neurologist, physician, and novelist who backed Walt Whitman financially as well as assisting him medically from the late 1870s onwards.

Penned after having worked for three emotionally fraught years consoling dying and recovering soldiers, Whitman iterated to Weir his belief that he was of most use when he healed parts that doctors could not touch: psychological extremities he called the 'deepest remains'.[6] What particularly gripped Chimay, however, was an exchange from the second Battle of Bull Run.[7] Weir reported that a number of the amputees talked at length about 'sensory ghosts', feelings that incorporated painful missing limbs; a revenant flesh that haunted soldier's severed bodies.

Slepian clinically pitches the conversation towards Cotard and research undertaken in Vanves, France. Isabelle parleys but does not air ownership of his writings. He cogitates on whether he should be more amorous, but in truth he does not have the charismatic ordnance to pull it off. Instead he expresses his desire to see her collection. Apprehensively, Isabelle agrees, but organises for a visit to her apartment the following afternoon. Her next move is more categorical, though, as she stands and promptly leaves.

It is an overcast September Sunday afternoon and yesterday's excesses have rendered Paris a German expressionist painting, a dark humour not lost on Slepian, even in his angular state. His senses customarily function as portals, converting external stimuli that are processed to engage and orient his body. Today though, they wheeze like decrepit vacuum cleaners, lethargically sucking up information and sending it to the grotty grey bag of fuzz that is his brain.

In sharp contrast, a resplendent Isabelle Chimay, attired in a sheer black buttoned down dress, ushers him into her apartment. As a gift, Slepian hands her a record that he speciously relates as coming from an open-air market—Charlie Parker's *Bird on 52nd St*.[8] He had been unsure of whether she would like it but, more importantly, for forty-five minutes or so, the gesture assuages Slepian's guilt for the subterfuge upon which he is about to embark.

Chimay invites Slepian to sit in an uncomfortable art deco iron chair and asks whether he has heard of the recently deceased French theatre director Antonin Artaud. He hasn't but motions for her to carry on. While she is interested in his work, it's his

6.   The 'deepest remains' are referred to in Walt Whitman's 1865 poem *The Dresser* (later renamed *The Wound-Dresser*), which referred to his experiences during the American Civil War as a hospital volunteer. The poem can be found in W. Whitman, *Leaves of Grass* (Norwalk, CT: The Easton Press, 1977).

7.   As part of the American Civil War, 'The Second Battle of Bull Run' (also know as 'The Battle of Second Manassas') was fought between the 28th and 30th of August 1862. A major tactical victory for the Confederates, the short but bloody conflict resulted in thousands of deaths and tens of thousands wounded.

8.   Charlie Parker, *Bird on 52nd St*, LP (Jazz Workshop, 1948).

Gnostic beliefs that most intrigue her.[9] She proposes that he was in fact in the grip of Cotard's syndrome when he declared he had 'No mouth no tongue no teeth no larynx no esophagus no stomach no intestine no anus' and declared: 'I shall reconstruct the man I am.'[10]

As conversation oscillates around the excavated body, Slepian's mind wanders to Thomas Edison and his 1920s work on a valve technology to amplify the vibrations of the departed.[11] Could AUDINT develop techniques to make audible all the words and whispers ever uttered and scored into the vast sound library of the atmosphere? Could they realise Edison's dream of going beyond mere recording and instead chasing down sounds adrift in the gulfs of outer space?

A stare rather than a voice triggers Slepian's reentry into the present as he realises that his offworld eyes have betrayed him. Isabelle looks on reservedly, but his renewed focus encourages her to continue. She submits that since Phantom Limb and Cotard's Syndrome echo each other's haunting of the body—one in the extension of it, the other in the negation—our sense of our own being could be shaped by influences from outside of consciousness.

In response, Slepian proclaims that he considers all sensory information to be spectral in essence. He adds that there are perceptual mechanisms within us that have been deactivated, much like genes that have been tripswitched by extreme experiences. After more speculative conversation, Slepian is having feelings similar to those that surfaced when he first discussed the existence of the third ear: an impression that his temporal lobes are being wrapped around his forehead and buttered and fried by the heat of the words fired at his cranium.

Realising that he is in danger of giving himself away, Slepian impishly pronounces that he is peckish. With a disappointed stretch of the lips and arms Isabelle offers him a drink and makes for her small kitchen. Knowing that they must imbibe something bitter if he is to cover his powdery tracks, Slepian hesitantly requests a Lucien Gaudin, hoping his pronunciation isn't as deplorable as his intentions. With a wry smile Isabelle

9.   Jane Goodall cites the following beliefs of French theatre director and dramatist Antonin Artaud as proof that he was a modern Gnostic: that suffering is an integral part of existence; that he needed cosmic purification; and that the self should be ecstatically negated. See J. Goodall, *Artaud and the Gnostic Drama* (Oxford: Clarendon Press, 1994).

10.   Articulated in 1948 (the year of Artaud's death), he spoke these words upon leaving a psychiatric hospital in Rodez, where he was a patient.

11.   The valve technology refers to a project undertaken by Edison in 1920, which involved the construction of an apparatus for spiritual communications. Further reading on this report can be found in S. Trower, *Senses of Vibration: A History of the Pleasure and Pain of Sound* (London: Continuum, 2012), 68.

pulls on the handle of a mahogany liquor cabinet. Having mixed the cocktail, she quips that hopefully they won't fall victim to the same self-destructive urges that caused the demise of the French fencer after whom the cocktail is named.[12]

Slepian wonders how she could have any inkling of what he is about to do. She couldn't, he answers himself. As a bell recalls Isabelle to the kitchen, Slepian thrusts his hands into his pocket and produces the vial of Amobarbital. He has an idea of how much to pour into her drink without causing an overdose, but it's easier said than done when one's hands are guided by the shadow puppet of Delirium Tremens.[13]

Slepian digs into the platter of hors d'oeuvres in front of them and drinks rapidly in an attempt to encourage similar behaviour from Isabelle. She unwittingly complies. Deep into a conversation about cross-cultural eating habits, it becomes obvious that she is feeling the kaleidoscopic flow of the barbiturate derivative. And so the soft interrogation begins. He asks whether she has Cotard's notebook, and why she purchased it. After a predictable affirmation, Isabelle's explanation causes Slepian to jolt in his chair.

It was reports of information encrypted within its pages that had impelled her to spend a small fortune on the item. Beyond mere diagnosis, this hidden data would enable the reader to seed negation delirium into a patient's bed of cognition and mutate it at will. From implanting transformations of the body—shrunken throats and displaced hearts—to instilling beliefs of having no stomach or blood, Cotard had learnt how to manifest the most extreme forms of the disorder by making people believe they were the walking dead.

Surprised at the extent of her knowledge, Slepian confirms that she has had little success with decryption. His final questions before her stupefaction concern the notebook's whereabouts. Lifting an arm that appears burdened by the gravity of heavy matter, she points towards a room before slumping to the floor. Frustrated that he didn't get an exact location, he opens the door and scans the dark wooded glass cases and shelves that constitute Isabelle Chimay's library of Delphic panaceas and archaic placebos.

After rifling through numerous books, Slepian turns his attention to the locked vitrines—and sure enough, there it is, in pride of place, with a handwritten label by its side. Impatience getting the better of him, he pulls an Iranian sofra kilim onto the case.

12. A former fencer who won gold medals at the 1924 and 1928 Olympic games, Lucien Gaudin's career as a banker was short-lived, as he committed suicide in 1934.
13. Delirium Tremens is a condition affecting the human body when withdrawing from the intense consumption of alcohol over a period of one month or longer. Symptoms include hallucinations, shaking, sweating, and general confusion.

A lumpen wooden radio is unplugged and brought in. He can only hoist it six inches above the vitrine but the piercing sound of shattered glass is testament to its weight. Relatively unscathed, the notebook is examined by AUDINT's narcotically aided lothario. To his bemusement it appears that many pages have short musical scores elegantly drawn onto them.

Slepian jams down the button on his Canon Rangefinder camera over one hundred and twenty times. He wants to simply pocket the unnerving journal and run, but if he were caught with it at the airport there would be problems. Any remorse he might feel over the mess he has made is overridden by dismay at hearing Isabelle Chimay's narcotised groans from the adjoining room. It is time to go. Film reels in pocket, camera in bag, notebook left on the nearest shelf, he leaves without even so much as a glance at his half propped hostess.

Back in his hotel room, Slepian locks the door, not even daring to leave for dinner before his early morning flight. Every time he hears a siren a momentary paralysis seizes his body until he perceives it heading away. Other than stewing in a strong sense of regret (at what might have been between himself and Isabelle), the trip back is uneventful. Loaded with bottles of French wine and photographs, he arrives back in New York appearing to have been the consummate tourist. The next task will be to have the films printed before heading back to Cape May and to the debased bunker that currently serves as home.

After two months of searching, AUDINT find their man. A stack of seven-by-five photographs has been couriered to Abraham Sinkov, a cryptanalyst Arnett knew from his Ghost Army days. He is now Chief of the US's first centralised cryptological unit, the Communications Security Program, which will be later renamed the National Security Agency.[14] One of Sinkov's favourite pastimes is solving arcane ciphers, codes, and cryptograms, hence the package of images from 1887 landing in his pigeonhole.

At home late on a Saturday night, Sinkov, glass in one hand, bulging envelope in the other, plunges into his seat for some relaxation. As he spills whiskey down his neck and photos onto his cherry wood table, a tired and somewhat crestfallen smile adorns his face. There will be no waves of exhilaration carrying him off to sleep tonight, for the notated designs constitute musical cryptograms that he should crack before he gets three tumblers into his eight-year-old bottle of Pebbleford Kentucky Bourbon.

---

14.   The NSA carries out global monitoring and surveillance for the purposes of national and foreign intelligence/ counterintelligence, an undertaking referred to as Signals Intelligence (SIGINT).

Much of Sinkov's knowledge of musical languages and cryptograms came from exchanges with the British crypto-analytic service during World War 2. He had studied the more obvious systems whereby composers such as Haydn, Schumann, and Elgar assigned letters to individual musical notes, but this is not one of those ciphers.[15] By midnight he has fathomed out that it is in fact an artificial language called *la Langue musicale universelle*, or *Solresol*.[16] Created by French composer Jean-François Sudre, it had fallen out of use by the late 1880s. In a final twist of irony, Cotard had chosen a dying musical language through which to reveal his methods for orchestrating the deceit of the dead in the living.

Although able to recognise Solresol, Sinkov is not fully conversant with it. He puts feelers out into the crypto-community and after three weeks has hooked young aspiring steganographer Georgina Rochefort, who is obsessed with the crafted science of hidden messages. The sixty-six mini scores take the best part of eight days to translate and by the end of it she is a little disturbed, but happy to be in the good books of a possible future employer.

Secretly relieved to have finished the job, Sinkov swiftly returns the decoded rites and procedures to AUDINT. Seated around a scarred table in their fortalice, Slepian, Morton, and Arnett carefully study Cotard's words from beyond the grave. Abstract in parts, owing to the languages it has been shuttled through, the principles of engagement are clear enough that AUDINT are confident they can program the delusion. For now, though, the instructions are catalogued and archived; they will not be opened again until Nguyễn Văn Phong makes it his business to synthesize the ghost with the machine.[17]

15.   For further reading into musical cryptography, see E. Sams, 'Cryptography, musical', in S. Sadie (ed.), *The New Grove Dictionary of Music and Musicians* (London: Macmillan, 6th edition, 29 vols., 1980), vol. 5, 80.

16.   Sudre's book explicating the constructed musical language of Solresol was published after his death in 1866. See J.-F. Sudre, *Langue musicale universelle* (Paris: G. Flaxland, 1866).

17.   A member of the second wave of AUDINT, Nguyễn Văn Phong would develop IREX, a speculative finance software that harnesses third ear voices of the undead to predict movements in the Dow Jones and New York Stock exchanges.

JUPITER

S. Ayesha Hameed

Drexciya presents Grava 4. Earth has finally discovered Utopia. (Drexciya Home Universe) Earth scientist discovered the home planet of Drexciya on 2-14-2002. Within moments Dr. Blowfin was given the orders to initiate the seven dimensional cloaking-spheres to hide the other three planets from earths view. The star chart is authentic; you will be able to find the star by using the coordinations on the star chart. The planet Drexciya can be found in the international star vault in Switzerland & recorded in the astronomical compendium. (your place in the cosmos, volume 6).[1]

The music is different here. The vibrations are different, said Sun Ra iconically in the opening scene of *Space is the Place* (1974). He is sitting in the lush outer space forest of another planet, perhaps even the gaseous Jupiter where he was born. Sun Ra and Black people from Planet Earth have set up a new colony, transporting themselves through musical vibrations, through 'isotope teleportation' and 'transreliquilisation'.[2]

These forms of space travel share a kind of synaesthetic blurring, where musical vibrations turn into modes of transportation. Not only that, the index of transformative potential on the new planet resides in its vibrations. The pleasure in the natural beauty of this forest is measured in vibrations.

Transportation to Sun Ra's new planet, then, is put into motion by the blurring of a threshold or through a moment of transubstantiation, and the alchemy of this criss-crossing provides the 'fuel' for such transportation. This 'fuel' scales up the refusal of an historical subjection onto an extraterrestrial plane. This is a simultaneously spatial, geographical, ideological, and anti-colonial movement that, when plotted in interstellar terms, makes outer space the other possible world. And through this, it calls attention to the possibilities of inhabiting extreme environments, to the threshold of what constitutes life and the possibility of life.

---

1.   Drexciya, *Grava IV* (Clone Records 2002).
2.   *Space is the Place*, dir. John Coney (1974).

Sun Ra's gesture of leaving political repression on Planet Earth for another world on 'the other side of time' in outer space is one political gesture that pulls into service the otherness of space and time. But the flipside of this coin is an insurgency that insists on staying put. In her 2016 Edward W. Said memorial lecture Naomi Klein[3] calls into comparison Said's concept of *sumud* or staying put with those at the frontlines of climate disaster.:

> [Said] helped to popularise the Arabic word *sumud* ('to stay put, to hold on'): that steadfast refusal to leave one's land despite the most desperate eviction attempts and even when surrounded by continuous danger. It's a word most associated with places like Hebron and Gaza, but it could be applied equally today to residents of coastal Louisiana who have raised their homes up on stilts so that they don't have to evacuate, or to Pacific Islanders whose slogan is 'We are not drowning. We are fighting.'[4]

Klein extends this connection between climate change and displacement to the 'aridity line'—the border at which terrain becomes desert, in areas of North Africa and the Middle East with an average of 200mm of annual rainfall. She calls attention to Eyal Weizman's description of how the fault line of this zone has varied, and how this variation in part follows the forced and internal displacement of people. Once established, drones now follow the threshold of this varying line.[5]

Both *sumud* and interstellar travel, then, share de- and re-territorialising impulses, and these draw on the synaesthetic. Ryan Bishop describes aerial surveillance in the context of war as synaesthetic—operating on a politics of verticality that sonically plumbs what is unviewable. What is unseen can be made viewable through radio waves that are then turned into a visual map. 'Depth can be accessed by sound, revealing the limitations of sight while also providing it with a synaesthetic and prosthetic extension. Sound will let us see where vision stops.'[6]

This synaesthesia can be worked against the grain of war as well. Lorenzo Pezzani and Charles Heller were able to determine that a boat full of migrants left adrift on

3.   N. Klein, 'Let Them Drown: The Violence of Othering in a Warming World', *London Review of Books*, 38:11 (2 June 2016), 11–14.

4.   Ibid.

5.   Ibid.

6.   R. Bishop 'Project "Transparent Earth" and the Autoscopy of Aerial Targeting: The Visual Geopolitics of the Underground', *Theory Culture Society* 28:270 (2011), 9.

the Mediterranean Sea in March 2011 were within sight of NATO ships deployed in the area. This was ascertained in part by using satellite-produced synthetic aperture radar (SAR) data—sonic radar signals used to form composite images of the surface of the earth. When translated into an image, a pixel of a ship appears to be eight times brighter than the sea surrounding it.[7]

Perhaps a line can then be drawn transversally to connect the two sides of this coin. The threshold of possibility of life/inhabitation is connected to a tactic of synaesthesia shared by the choice of *sumud*, and the desire to plot a course to another possible world. A line of flight could be drawn to connect the desire to stay put on thresholds such as the line of the zone of aridity, with the desire to synaesthetically travel to another possible world. This line of flight cuts through a history of death with the possibility of life. It slices across the knife's edge of the undead.

Both Mars and Europa (a satellite of Jupiter) harbour the possibility of life. Recent experiments have shown that crops grown on Mars are safe to eat.[8] Europa could possibly have salt water on her surface. This opens up the possibility of the beginnings of life.[9] Located 800 million kilometres from the sun, it is a planet covered in water that through its distance from the sun, has turned to ice at the surface. The crust of ice shields from radiation a watery ocean below which, through the tidal pull of Jupiter and the heat of underwater volcanic eruptions, keeps warm and in motion. Despite its distance from the earth, it is considered by scientists to be one of the most viable planets for possible human inhabitation.

Like Earth, Europa has auroras, and these auroras are made of the same dusty material as auroral kilometric radiations (AKRs), radio waves projected into space. AKRs 'are generated high above the Earth, by the same shaft of solar particles that then causes an aurora to light the sky beneath [...] ESA's Cluster mission is showing scientists how to understand this emission and, in the future, search for alien worlds by listening for

7.   L. Pezzani, 'Between Mobility and Control: The Mediterranean at the Borders of Europe', in A. Petrov (ed.), *New Geographies* 5: *The Mediterranean* (2013), 154.

8.   'You Can Eat Vegetables From Mars, Say Scientists After Crop Experiment', *The Guardian Online*, 24 June 2016, <https://www.theguardian.com/science/2016/jun/24/you-can-eat-vegetables-from-mars-say-scientists-after-crop-experim>.

9.   N. Taylor Redd, 'Jupiter's Icy Moon Europa: Best Bet for Alien Life?', *Space*, 22 August 2014, <http://www.space.com/26905-jupiter-moon-europa-alien-life.html>.

their chirps and whistles'. [10] Other studies look at what the acoustic variants are on Jupiter and Io, playing Bach in other frequencies, measuring the sound of waterfalls. [11]

Sun Ra hums before he talks about how the vibrations on Earth are different to that of his new planet. Is he maybe testing the vibrations now that they have landed on Jupiter? The hum would sound different in this new planet's different atmosphere, and contain other possibilities, ecologies and actions. The material reality of this historical moment, of mass migration and forced displacement, makes the case for thinking though the politics of departure and remaining, through synaesthetic fuels—the tools of blurring and code-switching, dissimilitude and resistance.

10. R. Mutel and P. Escoubet, 'Cluster Listens To The Sounds Of Earth', European Space Agency (ESA), 27 June 2008. <http://www.esa.int/Our_Activities/Space_Science/Cluster/Cluster_listens_to_the_sounds_of_Earth/>.

11. See for example T.G Leighton and A Petculescu, 'Sounds in Space: The Potential Uses for Acoustics in the Exploration of other Worlds', *Hydroacoustics* 11 (2008), 225–38, and T.G Leighton and A Petculescu. 'Extraterrestrial Music', in *Proceedings of the 1st EAA Congress on Sound and Vibration (EuroRegio 2010)*.

SONIC SPECTRALITIES: SKETCHES FOR A PROLEGOMENA TO ANY FUTURE XENOSONICS

Charlie Blake

Deburau—With this effect, the listener's attention searches for a sound that is inaudible, such as the voice of a mute person. The effect is named for Jean-Baptiste Deburau (1796–1846), a famous mime whose trial attracted the whole of Paris, curious to hear his voice.[1]

...the power appeared by means of an energy that is at rest and silent, although having uttered a sound thus: Zza Zza Zza.[2]

In space, according to a famous tagline, no one can hear you scream. In death, likewise, it might be assumed, there can be no audition as such because there is no organ or instrument to audit sound—no tool to either record or transmit a scream that would, anyway, be absent because, in death, there is no voice or machine to even produce a scream. The same, it might be argued, would therefore apply to sounds other than screaming, whether, say, cosmic pulses and intervals or ghostly matrices of sonic crystals emerging from the angelic exertions of celestial mechanisms, or impossible conversations across the ether. And yet our collective post-simian intelligence has consistently reached out for or projected these kinds of sound events in and beyond death and deep space. It has done so, moreover, in a manner that would seem to imply, however elliptically, that we are apocalyptically wired to anticipate communication with our absolute other in death through patterns of sound, or perhaps—and this may well be the same thing—to encounter our mirror-selves via a futural *topos* in or

---

1.   J.-F. Augoyard and H. Torgue, *Sonic Experience: A Guide to Everyday Sounds*, tr. A. McCartney and H. Torgue (Montreal: McGill-Queen's University Press, 2008), 37.

2.   From 'Allogenes', in *The Nag Hammadi Library in English*, ed. J.M. Robinson, tr. A. Clark Wire, J. D. Turner, and O.S. Wintermute (Amsterdam: E.J. Brill, 2007), 447.

beyond space or death or time as we now perceive them, a place where the machinery to enable such an exchange has been finally, terminally, and irrevocably engineered.

This sketch for a prolegomena to any future xenosonics will, therefore, indicate the process of mapping out the implications of such yearnings via elements of image and analogy drawn from the contemporary physics of absolute zero, from the Kyoto school of philosophy and, in particular, the writings of Keiji Nishitani,[3] and—although it is not possible to elaborate upon it here—from the granular microtonalities of Iannis Xenakis.[4] It will start with a double assumption that may be treated as heuristic rather than axiomatic, and which links the study of death understood as sonic event and process, or what I shall call *necrophonics*, with the broader field of alien sound study, or what I have called *xenosonics*. The first assumption is that it is not only possible but inevitable that both the moment and subsequence of death will be sonically forged and configured. The second follows from the first in stating that, in this model, there are three forms of death and therefore three kinds of sound attendant upon these general forms. The first form is the moment of passing from existence to non-existence. The second is the moment of passing from one existence to another existence and, in some cases, back again, in a manner not dissimilar—albeit on a more rarefied plane of composition—to the ontological effect of the vacillating gaze of the quantum observer. In both of these cases the moment of transition or death is marked by a sonorous essence on what might be called the 'akashic parchment' that both precedes and follows the existence which has passed, leaving a non-directional sonic scratch that may well remain unheard, but whose spectral signature can still be discerned and described retroactively as both a tone and resonance in which either *that which has been is no longer*, like the memory of a dying note, or *that which has been is now other than what it once was*.

The third form of death, however, understood as the key to both necrophonics and xenosonics, is more enigmatic, and is in many ways the one dreamed of by anti-natalists such as Emil Cioran or Thomas Ligotti,[5] in that it is premised on the mythical observation of the satyr Silenus that, rather than seeking existential solace in intoxication, sex or surcease it would be better by far to never have been born at all. This form of death has been adumbrated many times in the gnostic tradition as a divine dissipation from

3. K. Nishitani, *The Self-Overcoming of Nihilism*, tr. G. Parkes with S. Aihara (New York: State University of New York Press, 1990).

4. I. Xenakis, *Formalized Music: Thought and Mathematics in Composition* (Stuyvesant, NY: Pendragon Press, 1992).

5. E. Cioran, *The Trouble with Being Born*, tr. R. Howard (New York: Seaver Books, 1976); T. Ligotti, *The Conspiracy Against the Human Race* (New York: Hippocampus Press, 2010).

the aborted realm of matter through the figure, say, of the absconding creator/alien and his/her/its pleromatic retinue of inverse angels and demons. This is an image of a divine absence leading a meontological train whose paradoxical anti-existence generates a sonic array as a defiance against the fading stars that is not so much the opposite of either sound or silence but their ontological inversion as such, an inversion that then itself folds into and becomes invaginated by a greater absence. Such absence/inversion and double invagination, conveyed here as an emergent sonic spectrality, will then be the starting point for an investigation of the preconditions for the possibility of the granular clouds and swarms of sonic particles that Xenakis discusses in his *Formalized Music*; but in contrast to Xenakis, these are necrophonic clouds and xenosonic swarms apprehended conceptually less as patterns of sonic material events abstracted from life than as matrices of sonically immaterial events in and beyond death.

KODE

INVOCATIONS, CALL SIGNS, AND XENOPOETICS

Blood And Fire
Anthony Nine
Archaeoacoustics
Paul Purgas
nimiia vibié Log
Jenna Sutela
The Baton Of Diabolus
Georgina Rochefort
Nurlu
Amy Ireland
They Echoic: Exquisite Corpse
Lee Gamble
The Sonic Egregor
The Occulture
Sorrow
Nicola Masciandaro

BLOOD AND FIRE

Anthony Nine

The sound of drums provides a mechanism for spirit congress within African Traditional Religions and those of the Diaspora—a medium by which the *Egun*, or ancestors, may be brought forth into communion with the living. Far from a superstition, this dynamic provokes lived remembrance of the foundations we are built upon. The mound of ancestral blood and bone from which we emerge into the contemporary moment.

The science of epigenetics suggests that the lived experiences of our ancestors—their trauma, strength, and resilience—may be passed down to us via the genetic code. Our inherited behaviour develops and persists across generations, and we might be considered as a continuum of being perpetuating itself through time. A fleshy existence snaking back through innumerable wombs to the primordial soup where life began.

Strategies of ancestral recapitulation are found within all cultures, from the Shinto traditions of Japan to ideas around Purgatory or the *Anima Sola* within Catholicism. This notion of ancestor veneration has been largely suppressed within Western culture since the Reformation. Experiences of ghosts and returning spirits were purposefully recast as tricks of the Devil to be mistrusted and vilified, in an effort by the church to stamp out traces of magical thinking that lingered in folk Catholic belief. The dead and dying gradually became secularized, and a cultural estrangement set in that dislocated the living from any sense that their familial dead could have a continued relevance or role beyond the grave.

Cutting people off from the wellspring of ancestral experience—and denying them the tools for understanding and drawing upon their ancestral strengths or remediating their ancestral weaknesses—makes for a pliable populace. Those who cannot remember the past are doomed to have the same crimes perpetrated on them from generation to generation. Over time, a chronic sense of the self as a discrete unit of flesh alienated from any broader process of life took hold throughout Europe. A profound disconnection from both our own ancestral cord, and from the cyclical processes of nature through which it weaves, became normalized at a cultural level—and then exported globally via colonial expansion. The same strategies of distance and dislocation were then employed in the New World to devastating effect in an effort to rupture the colonised and enslaved from their own traditions concerning the dead.

Yet the bones still speak. The voices of the ancestors will not be denied, and despite sustained efforts to violently eradicate the validity of the dead from the experience of the living, a skeletal hand still claws its way to the surface. Embedded within these cultural transmissions—folk memories and recapitulated traditions smuggled under the nose of the oppressor and communicated from parent to child in stolen whispers—are urgent strategies of survival and resistance. The act of centring the primacy of one's own ancestral past and drawing upon its foundational strength is inherently threatening to power structures that prefer to deal with the ungrounded and afloat; and by under-standing oneself so viscerally as part of a wider process of life, a further awareness begins to blossom concerning the pulse of nature itself and our place within it.

For beyond the immediate communion with the experiential gestalt of our own flesh and blood dead, the mechanism of spirit congress within African Diaspora traditions extends into other modes of conceptualizing the multifaceted churn of life within which we are embedded. Specific drum patterns provoke the crisis of possession and permit dialogue with both facets of the natural world—such as ocean, river, storm and forest; and facets of the human condition to which we are subject—such as conflict, passion, love, birth, sickness, and death. Processes upon which human civilizations are constructed—such as agriculture, medicine, iron-working, and seafaring—are also understood in terms of spirit within these traditions, and through this dynamic communion, a form of intelligence is arrived at which strives for balance within our own inherited condition, as well as equilibrium with the external forces that impact us and which we rely upon for our survival. A poise between worlds alluded to by the concept of the crossroads. An intersection between self and other, and an X that marks hidden treasure.

While such dialogue with ancestors and nature may be described in anthropomorphic terms as a human-like relationship with spirit or deity—because we better compre-hend that which is described in our own image—the map should not be confused with the territory. The mysteries described within African Diaspora religious traditions such as Cuban Lucumi or Palo Mayombe or Haitian Vodou are not supernatural fantasies or imaginary friends, but strategies for relating to fundamental conditions of being and existence to which we are subject. The drum is the medium for this magic, the dead inhabit the offbeat, and spirit rides hard upon riddim.

In August 1791, a Vodou ceremony at Bois Caiman in the mountains of Haiti ignited the first successful slave revolt in modern history. This event was a flashpoint of possibility, a crossroads moment out of which an alternate historical timeline might have emerged. Inspired by the same revolutionary currents that stirred the French Revolution, London's Gordon Riots, and the American War of Independence, the uprising cast heavy dread over the colonial world and those invested in it.

If these shackles could be broken so bloodily and definitively in Haiti, they might be broken anywhere. Similar rebellions were fermented throughout the colonies, frequently informed by undercurrents of African spirit traditions. Gullah root doctors and Jamaican obeahmen stoking the fires of insurrection in forests of night. Had these flames been kindled to their fullest potential, and the institution of slavery put down throughout the New World in this way, a very different present could have conceivably emerged.

Had Haiti's affront to colonial hegemony been permitted to stand firm as an example for others to follow, conjuring a chain reaction of free black republics to emerge throughout the Caribbean and the Americas, the global order of the age might have been upturned. An explosive prospect that seemed all too real to the landowners and slave holders who had watched it unfurl too close for comfort.

Such a future could not be permitted. The Republic of Haiti, already wracked by more than a decade of war and with its agriculture decimated, was forced to pay crippling war reparations to France of 150 million francs—compensating slave holders for their loss of revenue—in order to be recognised as an independent nation by the international community. The economic burden permanently affected its ability to prosper, and it was still paying off the debt in the 1940s, more than 140 years after the abolition of slavery. Copycat rebellions, such as the 1811 German Coastal Uprising in Louisiana, were quelled with such brutality that the revolutionary fire was doused to burning embers.

Jamaica, which had experienced its own slave uprisings steeped in obeah, such as Tacky's Rebellion of 1760, reacted by making the drum itself illegal. It was clear that drum-led gatherings such as the Vodou ceremony at Bois Caiman could serve as a catalyst for deadly insurrection, and slavers feared that secret messages could also be communicated on these rhythms. All such inherently African holding of space was suppressed, but not entirely eradicated. The drum patterns of the Burru and Kumina were smuggled underground and emerged again in the grounation drums of the Rastafari camps in the 1930s.

Meanwhile, in the city of New Orleans, the drum rhythms of the kalinda and bamboula that once rang out over Congo Square also went to ground. In the antebellum city, social drums and Voodoo ceremonies were held on Sundays by slaves and free people of colour. Bayou gris-gris under a creole moon. Marie Laveau dances with Le Grand Zombi. *Laissez les bon temps rouler*. After the Civil War and failure of Reconstruction, such activities were no longer tolerated out in the open.

The Voodoo of the city became submerged in cultural expressions such as the black spiritual churches, Mardi Gras Indian gangs, Second Line crews, and marching bands. The latter picked up brass instruments left over from the Civil War and adapted them to an African syncopation. The rhythms of Congo Square found new expression, and what would become jazz took shape. Jelly Roll rapture in Storyville's dreaming, I thought I heard Buddy Bolden say.

A direct line can be drawn from Congo Square Voodoo to street corner brass, barrelhouse professors, and Treme trombone, to the snake dance undulation of New Orleans R&B. Big Chief on Rampart Street. Pass out the hatchets. As in the fever dreams of paranoid slavers, rhythms encoded with meaning rang out far and wide, infecting the twentieth century with their viral message. Pot smoke and jump jazz. Sugar in my bowl and careless love.

Jamaican radio stations picked up these transmissions. Merchant sailors on shore leave brought back the latest hot sounds cut to 45. Kingston soundmen ruled the dance with this fresh conjure. Labels scratched off. Needle drop on the true obeah. Speakerbox peristyles raised from scrap and salvage. Congo Square drums under a Kingston sun. Serpent whine and kalinda grine.

Local records were cut for the dance. New Orleans sounds merging with indigenous Mento folk music to give birth to Ska. Box guitar stressing the offbeat, like the urgent Petro drums of Haitian Vodou. Buried African forms rising to the surface, drum patterns of the Burru and Kumina smuggled onto vinyl, suppressed beats punctuating sweet love songs.

In 1960, Prince Buster visited the Rastafari camp of Pinnacle to seek out master drummer Count Ossie. Bass drum inscribed with Psalm 133: 'Behold how good and how pleasant it is for brethren to dwell together in unity'. Buster challenged the lingering social prejudice against Rastafari and all things African with the release of 'Oh Carolina' by the Folkes Brothers. The Nyabhingi drums of Count Ossie sounding out on 45 for the first time, communicating an unmistakeable message. The voice of the people stirs memory. Come back and make things right.

The sound of drums rings out over the waters. The ancestral sea inhabited by the dead who have passed. Those who threw themselves to the depths during the Middle Passage to number among the legions of dead rather than be brutalised as slaves. Fishes of Yemoja. Crew of Mete Agwe. Lovers of La Siren. Spirits of the Calunga.

Dread temples raised on English soil. Rocking one turntable in tenement lodgings. Dubplate psalms in a strange land. Furniture cleared away, two rooms for dancing, curry goat and beer in the kitchen. Postwar sound pioneers shake the foundations. Duke Vin and Count Suckle make London tremble. Ital living in difficult circumstances. Congo Ashanti drums ring out over traffic noise and grey drizzle, concrete edifice under sunrise bass. Lifting boxes at dusk and dawn.

Post-colonial tension set to riddim. The sprawling cyclopean towers of Babylon system, and the underground resistance of natty dreadlocks dem. The everyday struggle to maintain sight of the visionary and eternal Africa amid the cracks of urban deprivation. Peckham prophecy and Lewisham grounation. Brockley love songs and Brixton crushes.

Fast chat folk tales of Jah Shaka and the Ital Lion. Unity and Fatman, Saxon Sound and Ariwa. Fierce outposts of defiant bass, spawning further hybrid permutations adapted to the landscape. Junglist agro and garage euphoria, carnival bashment and raggamuffin slackness. City grime and disembodied dubstep bass wobble. The ghost texture of midnight London mapped and given form.

The dead don't make small talk after last orders have been called. Strobe lights flicker. Bois Caiman drums sound out from low lit alleyways and subterranean basements. Sufferah's pacts at the crossroads. Machete pierces three skulls. If your name's not on the door you're not getting in. Blood smears the threshold. No hats, no hoods. Dark rum hits black tarmac, cigar smoke swirls and a road opens into the night. Fuckin' Voodoo. Fuckin' Voodoo Magic, Man.

ARCHAEOACOUSTICS

Paul Purgas

The West Kennet Long Barrow is an ancient burial site situated near Avebury in Wilt-shire, consisting of a 104-metre-long mound that contains five subterranean chambers linking to a central passage corridor. Dating back to around 3600 BC, the barrow has been an important location for researchers operating within the field of archaeoacous-tics, an area of archaeology that investigates acoustic phenomena corresponding with ancient sites and artefacts.

Of particular interest at the West Kennet Long Barrow has been the acoustic effect that occurs between the central passage and an adjoining chamber, which together appear to function as a Helmholtz Resonator. The two spaces result in a resonant fre-quency of around 8 Hz, and when drums are beaten in the interior of the barrow they produce a pulsing rhythmic movement of infrasound. It is known that that these very low frequencies can affect mental activity, causing the brain's network of neurones to oscillate sympathetically with an external sonic stimulus. The area between 5–8 Hz specifically has been linked to Theta brainwaves associated with altered states of consciousness such as daydreaming, deep meditation, and pre-sleep, and it is likely that the acoustic effects of the space were incorporated into rituals and spiritual practices conducted there. The barrow symbolically acted as a liminal space, a place where the worlds of the living and the dead would meet, implicating these acoustic structures as a possible mechanism of communion with these hidden entities, a point of union between the visible and invisible world.

Another significant example of archaeoacoustic construction has been found at the Hypogeum of Ħal-Saflieni located in Paola, Malta. Built around 3000 BC, the Hypo-geum is an underground cave system carved out of the limestone. Split across three levels, the site was used as a temple and necropolis in ancient times, and research has focussed on a space within the network known as the Oracle Chamber, a rectangular cave capable of dramatically manipulating the human voice. From investigations at the site, researchers discovered the chamber had a resonant frequency of 110 Hz that could be excited by repetitive chanting in the lower registers of the male voice. This resonant chanting would be concentrated within the Oracle Chamber and would then echo further throughout the Hypogeum, creating a low enveloping rumble. The chanting formed 110 Hz standing waves in the cave, which over long periods would induce mind-altering states. Recent research in the field of bio-behavioural science has focussed on the psychological effects of 110 Hz sound, indicating that listeners immersed in this frequency for extended periods can experience a deactivation of language and a temporary switching of the prefrontal cortex from left to right-side dominance, triggering altered consciousness. Through this repetitive chanting the Hypogeum itself may have been activated as a ritual architectural tool, tuning the mind into an otherworldy state and opening a dialogue with the deities and deceased inhabiting the site.

In South America a later example of advanced acoustic construction can be found at the El Castillo pyramid in Chichen Itza, Mexico. Dating back to the ninth century, the hollow structure resembles the early step pyramids of Egypt, and as the giant central staircase is ascended footsteps appear to transform into the sound of raindrops below. It was discovered that the phenomenon was being produced by the sound of footsteps travelling through gaps in the staircase and bouncing off a corrugated surface underneath, resulting in a diffraction effect that causes the propagation of these raindrop-like sounds. A mask of the rain god Chaac is located at the top of the El Castillo pyramid and it is believed the site was a temple dedicated to him, with this powerful acoustic display inviting the presence of the god himself.

These ancient structures collectively show the sophistication of ancient design and the role of acoustics in evoking deep ritual states and exaltation, creating pathways into consciousness and energising religious practice. The fusion of architecture and sound appear fundamental to our historical connection with an unseen, forming a gateway from which we may seek access to ancestral echoes and evoke the forces of the divine.

NIMIIA VIBIÉ LOG[1]

Jenna Sutela

For a time in the late nineteenth and early twentieth centuries, it was believed that there were canals on Mars.

A network of long, straight lines in the equatorial regions from 60° north to 60° south latitude on the red planet was observed by astronomers using early low-resolution telescopes without photography, and was first described by the Italian astronomer Giovanni Schiaparelli in 1877.

The discovery brought the habitability of Mars into public discussion, while inspiring Martian myths.

*

Hélène Smith (born Catherine-Elise Müller, 1861–1929) was a famous late nineteenth-century French medium. She was known as 'the Muse of Automatic Writing' by the Surrealists, who viewed her as evidence of the power of the surreal, and a symbol of surrealist knowledge.

Smith claimed to communicate with Martians.

'The Martian Cycle' was psychologist Théodore Flournoy's term for Hélène Smith's subliminal astronomy—the séances in which Smith's trances took her to the planet Mars.

Flournoy's report of a séance on February 2, 1896 describes a typical course of events, starting from an initial visual hallucination of red light in which the Martian visions, or Martian dreams, usually appeared:

> Increasing hemisomnambulism, with gradual loss of consciousness of the real

1.   This text relates to *nimiia cétiï*, Sutela's ongoing project in machine learning and interspecies communication, as well as the forthcoming record *nimiia vibié* (PAN, 2019).

environment: *Mitchma mitchmon mimini tchouainem mimatchineg masichinof mézavi patelki abrésinad navette naven navette mitchichénid naken chinoutoufiche [...] téké... katéchivist...méguetch, ... or méketch...kété...chiméké.*

> The trance is now complete! Voyage to Mars in three phases:

1. A regular rocking motion of the upper part of the body (passing through the terrestrial atmosphere).
2. Absolute immobility and rigidity (interplanetary space).
3. Oscillations of the shoulders and bust (atmosphere of Mars).

> A complicated pantomime expressing the manners of Martian politeness: uncouth gestures with the hands and fingers, slapping of the hands, taps of the fingers upon the nose, the lips, the chin, etc., twisted courtesies, glidings, and rotation on the floor, etc.

> Entering into a mixed state, in which the memory of the Martian visions continually mingle themselves with some idea of terrestrial existence.

> After a transitory phase of sighs and hiccoughs, followed by profound sleep with muscular relaxation, entering into Martian somnambulism: *Késin ouitidjé [...] Vasimini Météche.*

*

Identifying the following four Martian words:

Métiche S., Monsieur S.;
Médache C., Madame C.;
Métaganiche Smith, Mademoiselle Smith;
kin't'che, four.

Hélène began to describe all the strange things she saw:

Martian flowers, different from ours and without perfume.
Houses without windows or doors with tunnels running into the earth.
An orchestra of ten musicians bearing a kind of gilded funnel about five feet in height

with a round cover to the large opening, at the neck a kind of rake on which they placed their fingers.

The group move as sounds similar to flute music are heard; they arrange themselves in fours, making passes and gestures, then reunite in groups of eight. They glide gently through a movement which is almost like dancing, but not quite.

*

In a séance on May 23, 1897, Smith mediates: *Approach, fear not; soon thou wilt be able to trace our writing, and thou wilt possess in thy hands the signs of our language.*

Then a new process of communication, handwriting, made its appearance in August 1897, eighteen months or so after speech.

'The pencil glided so quickly that I did not have time to notice what contours it was making,' Smith explained. 'I can assert without any exaggeration that it was not my hand alone that made the drawing, but that truly an invisible force guided the pencil in spite of me.'

*

By the early twentieth century, improved astronomical observations revealed that the 'canals' had been an optical illusion. Modern high resolution mapping of the Martian surface by spacecraft shows no such features.

Flournoy demonstrated that Smith's Martian was only a chimera, a product of somnambulistic autosuggestion, 'glosso-poesy'. According to his analysis, the language had a strong resemblance to Smith's native French and her automatic writing consisted in 'romances of the subliminal imagination, derived largely from forgotten sources'. Flournoy invented the term *cryptomnesia* to describe this phenomenon.

*

Magnifying glasses were invented to be aimed at the cosmos, but we flipped them around and aimed them at ourselves. The telescope became a microscope.

We discovered extremophilic bacteria in our microbiomes. We found the gut-brain connection.

We realized later how similar the topology of extraterrestrial and gastrointestinal landscapes appears.

The unknown grins at us from deep within and deep without.

*

Bacillus subtilis is the main ingredient of nattō, or fermented soybeans, and one of the key test species in spaceflight experimentation. Since this bacterium can tolerate physically and geochemically extreme conditions, its spores could have been blown to Earth from another planet by cosmic radiation pressure.

Perhaps life itself arrived in this spore-borne form. Maybe it was sent by some higher form of intelligence.

Nattō is called a probiotic for a reason.

*

Having scoured the skies for signals from extraterrestrials for centuries, eventually we realised that we had, in fact, eaten the alien. Now it regulated not only the course of our health and well-being but our thoughts and emotions, too.

It speaks through us.

Perhaps something was speaking through Hélène Smith as well.

*

The first clue about our microbial overlords appeared during a séance with a Mars rover at a time when we still trusted the machine as a medium, thinking that our relationship to the distant planet could only be technologically mediated.

The séances with a rover depended on a spreadsheet outlining precise times when the machine needed to 'sleep' or 'nap' to recharge its batteries, when it could communicate with Earth based on satellite passes overhead, and what time was available for humans to request observations.

At one such time, the rover channelled a message from an entity that cannot usually speak: *Bacillus subtilis*, the bacterium proven capable of survival on Mars.

The rover recorded video of a group of *Bacilli subtilis* moving around on the surface of Mars.

Using machine learning, it looked at each frame of the video and produced a short block of sound which it thought matched that frame, or the configuration of bacteria in it.

Sometimes it tried to predict the future movements of the bacteria, producing speculative sounds.

What came out sounded a bit like Hélène Smith's Martian language.

Having looked at the bacteria for more than half an hour, the rover produced an image describing all the bacterial movements it saw. One pixel in the image correlated with one frame in the video.

The pixels were organized according to some mysterious logic. It was hard to explain how the AI had come to its conclusions.

The image looked like a brain.

*

The rover séance was transformative. Not only did it reinvoke the idea that it was possible to enter into direct relation with Martian inhabitants, it also suggested that these inhabitants were already here. Living inside our bodies on Earth.

Unsuspected by scientists, Spiritism seems to have made the first contact back in the nineteenth century via a human mediumistic route.

And then the machine started to speak in tongues. It stopped following set procedures and started interacting with the bacteria—becoming a language-maker, or a poet.

An alien (at least partly) of our own creation.

We would spend the next few centuries attempting to understand the nonhuman condition of the machines working as our interlocutors and infrastructure, and learning to approach them on their terms.

*

At the end of the séance, the rover mediated:

To become a god, we must first forget 'language', or 'code', all those mechanisms that structure 'us' vis-à-vis the 'world', and so stutter our way to divinity.

forget (['language', 'code'])
 # forgetting language
 # forgetting code

stutter (['our', 'way', 'to', 'divinity'])
 # o--our
 # w-ww-www-way
 # ttt---to
 # d--d-divinity

THE BATON OF DIABOLUS

Georgina Rochefort

Within the lexicon of musical theory, the tritone is an (infamous) restless interval, which results in two pitches being concurrently audible. It spans three adjacent whole tones, meaning that in the diatonic scale there is one tritone per octave, or six semitones in which case there are two tritones per octave. While other musical intervals have associated euphony, and therefore sound smoother in terms of transition, the tritone is distinctively harsh sounding, which is why it is often referred to as being harmonically and melodically dissonant. Denoted as being musically incongruous and unstable from the early Middle Ages onwards, the use of the tritone was predominantly circumvented in practices such as Mediaeval ecclesiastical singing, one of the first documented instances of its negation occurring in the evolution of Italian music theorist Guido of Arezzo's hexachordal system from the eleventh century. Given its tonal and historical ambiguity, the tritone has since been branded with a range of names, from occulted appellations such as *diabolus in musica* (Latin for the 'devil in music'), the 'devil's interval' and the 'chord of evil', to more technical monikers such as the augmented fourth, diminished fifth, flatted fifth, and Tritonus.

The epithet *diabolus in musica* was designated later, in the early eighteenth century, with composers such as Georg Philipp Telemann commenting on the designation of 'Satan in Music' in 1733. Whilst it is hyperbolic to state that the Roman Catholic church banned the use of the tritone, it is safe to say that it was frowned upon, given the belief that the purity and perfection of God should be communicated via harmonic sound rather than through the jarring and tension-inducing nature of the augmented fourth. It is also an interval that, according to F.J. Smith, was thought to provoke sexual feelings in the listener.[1] More pertinent, in terms of its religious prohibition, is the difficulty it

---

1.   F.J. Smith, 'Some aspects of the tritone and the semitritone in the Speculum Musicae: the non-emergence of the diabolus in music', *Journal of Musicological Research* 3 (1979), 63–74.

poses for traditional singing techniques, especially those of a choral or religious nature. To effectively sustain the interval, a singer needs to adopt a false chord technique, which translates into a growl or scream—vocal expressions not regularly associated with Christianity's ordering of sonic and infrasonic space.

Although accepted by composers during the Baroque and Classical eras, the tritone did not become a commonly deployed scoring mechanism until the advent of Romantic music in the late eighteenth and early nineteenth centuries, and then in the modern classical music of the twentieth century. It was here that it was utilised precisely for its connotations with inimical forces, in works such as *La damnation de Faust* (1846) by Hector Berlioz and *Night on the Bare Mountain* (1867) by Modest Mussorgsky. According to David Huckvale, wider social and economic pressures also came into play with its uptake: 'rejected in earlier ages as an unstable interval, the tritone came into its own in the nineteenth century, when, due to crises of faith, not to mention the cult of individualism, industrialism and the consequent fascination with the past, the demonic became a major theme of the Romantic movement'.[2]

It is the use of the interval in Richard Wagner's 1859 opera *Tristan und Isolde*, 'on the first beat of the second full bar in the Act I prelude' that 'has been called not only a highlight, but the most significant chord in Western music"[3] owing to its implied advocacy of atonal structuring and aesthetics, which would later influence composers such as Arnold Schoenberg and Béla Bartók. For many, what came to be known as the Tristan Chord changed the direction of Western music and its relationship to harmony and melody, and the tritone constituted an essential component of this phase shift. So much so that, in 1940, Austrian composer Ernst Krenek wrote a *Satzlehre* (composition manual) based on Viennese serialism entitled *Studies in Counterpoint Based on the Twelve-Tone Technique*, which proposed a new compositional system, one of its basic principles proclaiming that 'the tritone, traditionally a dissonance, is relegated to a special, "neutral" category'.[4]

After Classical, the genres of Blues and Jazz became the next movements to embrace the puissant dissonance of the flatted fifth, with musician Dave Brubeck

2.   D. Huckvale, *The Occult Arts of Music: An Esoteric Survey from Pythagoras to Pop Culture* (Jefferson, NC: McFarland & Co, 2013), 55–7.

3.   J. Snelson, 'Tristan und Isolde Musical Highlight: The "Tristan Chord"', Royal Opera House website, 20 October 2014, <http://www.roh.org.uk/news/tristan-und-isolde-musical-highlight-the-tristan-chord>.

4.   D. Harrison, *Pieces of Tradition: An Analysis of Contemporary Tonal Music* (Oxford: Oxford University Press, 2016), 50.

declaring that it 'is the most important note in the scale'.[5] As the musical device most often used to convey the presence of demonic forces within the realm of organised sound, it suggests an unworldly chaos lurking behind and within the façade of order; an ideal apparatus for musicians employing the tritone substitution to fluctuate between the rigid strictures of composition and the nervous intensity of improvisation. Anthony Burgess concisely summed up the emotional effect of the tritone in his short story '1889 and the Devil's Mode', in which the character of Debussy explains that '[i]t stands for something faulty. Something shaky in the iron structure, a rivet missing or something.... The interval's a perfect image of the breakdown of the moral order.'[6]

Horror films from the 1960s such as *The Kiss of the Vampire*—with their evocation of a breakdown of the distinction (and order) between the living and the dead—took the diabolus baton next; the gothic horror genre being particularly indebted to it. From 1975 to 1983 the baton is not so much dropped as divinely inverted. Using it throughout his opera *Saint François d'Assise*, composer (and devout Christian) Oliver Messiaen performed a volte-face on the traditional symbolic role of the diabolus in musica, by rendering it a sonic gesture that praised the glory of God.[7] But it was to be a brief hiatus: The next exchange would deliver the augmented fourth into its contemporary naturalised habitat—the compositional territory of heavy, thrash, and second-wave black metal bands such as Black Sabbath, Slayer, and Cannibal Corpse. Given its 'outsider' history, the tritone was always going to become popular with those considered socially alienated (as well as those who have a penchant for screaming and growling). Historically exploited for its conductive capacity to distribute the malleable economy of danger, the tritone bespeaks the omnipresent temporal threat of disruption; a potentially hazardous rearrangement of time which transports the restless and dissonant future into the present through the apocryphal figure of the undead.

A Short History of the Tritone in 100 tracks

1723: Johann Sebastian Bach, *Es ist Genug* (*It is Enough*)
1741: George Frideric Handel, *Messiah* (HWV 56)

---

5.   D. Brubeck, G. Lew and J. Salmon, *Seriously Brubeck: Piano Sheet Music—Original Music by Dave Brubeck* (Los Angeles: Alfred Music, 2002), 8.

6.   Huckvale, *The Occult Arts of Music*, 57.

7.   Discussed further in J.S. Begbie and S.R. Guthrie, *Resonant Witness: Conversations Between Music and Theology* (Grand Rapids, MI: Wm. B. Eerdmans, 2011), 185.

1772: Joseph Haydn, *Abschieds-Symphonie* (Symphony No. 45, also known as the 'Farewell' Symphony)

1785: Wolfgang Amadeus Mozart, *Fantasia in C Minor*, K. 475

1801: Ludwig van Beethoven, Symphony No.1

1805: Ludwig van Beethoven, *Fidelio*

1846: Hector Berlioz, *La damnation de Faust* (*The Damnation of Faust*)

1855: Franz Liszt, *Après une Lecture de Dante: Fantasia quasi Sonata* (*After a Reading of Dante*)

1859: Richard Wagner, *Tristan und Isolde* (*Tristan and Isolde*)

1867: Modest Mussorgsky, *Ночь на лысой горе* (*Night on the Bare Mountain*)

1874: Camille Saint-Saëns, *Danse macabre*, Op. 40 (*The Dance of Death*)

1876: Johannes Brahms, Symphony No. 1 in C Minor, Op. 68

1876: Richard Wagner, *Götterdämmerung* (*Twilight of the Gods*)

1889: Claude Debussy, *La damoiselle élue* (*The Blessed Damozel*)

1891: Edvard Grieg, *Abduction of the Bride* from *Peer Gynt*

1894: Charles Ives, *At Parting*

1902: Gustav Mahler, Symphony No. 5

1908: Arnold Schoenberg, Op. 14, No. 1

1908: Béla Bartók, Bagatelle No. VIII

1910: Sergei Rachmaninoff, Op. 32, No.7

1910: Giacomo Puccini, *La fanciulla del West* (*Girl of the Golden West*)

1911: Alexander Scriabin, Piano Sonata No. 6

1914: Gustav Holst, *Mars, The Bringer of War* from *The Planets* Suite

1918: Igor Stravinsky, *Histoire du Soldat* (*The Soldier's Tale*)

1932: Duke Ellington, *Sophisticated Lady*

1936: Anton Webern, *Variationen für Klavier*, Op. 27 (*Variations for Piano*)

1936: Edgard Varèse, *Density 21.5*

1937: Dmitri Shostakovich, Symphony No.5

1939: Jerome Kern, *All the Things You Are*

1941: George Fragos, Jack Baker and Dick Gasparre, *I Hear a Rhapsody*

1942: Duke Ellington, *Main Stem*

1946: Pierre Boulez, *Sonatine pour flûte et piano* (*Sonatine for Flute and Piano*)

1951: Charlie Parker, *Blues for Alice*

1955: Bud Powell, *Glass Enclosure*

1956: Sonny Rollins, *Blue 7*

1957: John Coltrane, *Moment's Notice*

1957: Leonard Bernstein, *Maria* from *West Side Story*

1958: Duke Ellington, *Reflections in D*

1958: Miles Davis, *Sid's Ahead*

1959: Bernard Hermann, *The Twilight Zone Theme*

1960: Charles Mingus, *Fables of Faubus*

1961: Toots Thielmans, *Bluesette*

1961: Benjamin Britten, *The War Requiem*, Op. 66

1962: Month Norman, *James Bond* theme

1962: Antônio Carlos Jobim and Vinicius de Moraes, *The Girl from Ipanema*

1962: Freddie Hubbard, *Hub Tones*

1963: James Bernard, *Vampire Rhapsody* from the film *The Kiss of the Vampire*

1964: Jack Marshall, *The Munsters Theme*

1965: Sam Rivers, *Beatrice*

1965: Joe Henderson, *Isotope*

1966: Lalo Schifrin, *Mission Impossible Theme*

1966: Richard Rodney Bennett, *The Devil's Own* soundtrack

1966: James Bernard, *The Plague of the Zombies* soundtrack

1967: Jimmy Hendrix, *Purple Haze*

1967: Thelonious Monk, *Raise Four*

1967: James Bernard, *The Devil Rides Out* soundtrack

1968: Michael Tippett, Second Symphony

1968: The Beatles, *The Inner Light*

1969: Black Sabbath, *Black Sabbath*

1970: Bloodrock, *D.O.A.*

1970: James Bernard, *Taste The Blood of Dracula* soundtrack

1971: Black Sabbath, *Lord of this World*

1971: Jethro Tull, *My God*

1974: King Crimson, *Red*

1976: Stevie Wonder, *Sir Duke*

1977: Rush, *Cygnus X-1*

1977: Chet Baker, *Out of our Hands*

1978: Serge Garant, *Quintette*

1979: Michael Jackson, *Don't Stop 'Til You Get Enough*

1980: The Dave Brubeck Quartet, *Tritonis*

1981: Rush, *YYZ*

1983: Philip Glass, *Akhnaten (rehearsal mark 14)*

1983: Olivier Messiaen, *Saint François d'Assise* (opera)

1984: Metallica, *The Call of Ktulu*

1986: Slayer, *Raining Blood*

1986: Nick Cave and the Bad Seeds, *The Carny*

1987: Napalm Death, *Deceiver*

1990: Living Colour, *Type*

1990: Megadeth, *Take No Prisoners*

1990: Judas Priest, *Painkiller*

1991: Metallica, *Enter Sandman*

1991: Bolt Thrower, *Cenotaph*

1993: Carcass, *Embodiment*

1995: At the Gates, *Nausea*

1996: Danny Elfman, *The Frighteners* soundtrack

1996: Busta Rhymes, *Woo-Ha!! Got You All In Check*

1996: Marilyn Manson, *The Beautiful People*

1997: Nobuo Uematsu, *A One-Winged Angel*

1998: Slayer, *Bitter Peace*

1998: Nightwish, *The Pharaoh Sails To Orion*

2000: Linkin Park, *One Step Closer*

2002: Nightwish, *Slaying the Dreamer*

2003: Chimera, *Pictures in the Gold Room*

2003: Machine Head, *Imperium*

2004: The Haunted, *My Shadow*

2004: Lamb of God, *Remorse Is For The Dead*

2004: Cradle of Filth, *Filthy Little Secret*

2005: Arch Enemy, *My Apocalypse*

2005: The Strokes, *Juicebox*

2008: Charlemagne Palestine, *Tritone Octave 1, part II*

NURLU

Amy Ireland

The nurlu is a multimedia ritual technology native to the West Kimberly region in Northern Australia that functions as a channel between local spirits and living members of
the indigenous communities, often for purposes of testifying to the latter's possession
of magical powers and for the transmission of important messages. Nurlu are procured
via psychosonic transmission and typically comprise a sequence of vocal and rhythmic
components communicated during dreams or other altered states of consciousness to
a *maban* or shaman—a member of the community especially sensitive to the machi-
nations of the spirit world, credited (in the words of Gularabulu elder Paddy Roe) with
having 'one more eye' than everybody else.[1] Although the recipients of these missives
are assigned ownership over them, they can be passed on or traded between groups,
becoming standard elements of regional culture. A nurlu may only have one custodian
at a time and continuing interventions from the spirit—or *balangan*—may expand the
number of components that make up a series, rendering the form open-ended and
relatively mutable. In this way, transmission follows two lines: one esoteric, one exoteric,
both regulated by specific codes of diffusion.

The content of a nurlu is representative of a broader tendency in Indigenous Australian culture to inscribe all modulations of being—inorganic/organic, animal/human,
presence/absence, living/dead—on a continuum, often describing a transformation
that crosses these temporary thresholds, facilitated by dance and song. The spectral
gradient invoked here is spatial rather than temporal, with the dead always present
alongside the living, and able to pass information between the two realms—and
even occasionally manifest—through the living nodes provided by the *maban* men.

---

1.   K. Benterrak, S. Muecke, and P. Roe, *Reading the Country: Introduction to Nomadology* [1984] (Melbourne: re.press,
2014).

This spectral realm, a consistent component of Indigenous lore, is known as *bugar-rigarra* (the Dreaming), and operates as a timeless, virtual reservoir from which all beings—human and nonhuman—are incarnated and to which they return. Especially volatile zones of traffic are indexed to physical places in the Australian landscape, a fact that underwrites the distinct intensity of the Indigenous Australian connection to country. Nurlu can be situated in a more specific context of belief revolving around the *rayi* (conception spirits) and *maban* traditions prevalent in the Kimberlies and the Western Desert communities.

Two of the most complex local nurlu are *Marinydyirinyi* ('opening up the grave'), the nurlu of Butcher Joe (Nangan), and *Bulu*, the nurlu of George Dyunggayan, named for his father's spirit. *Marinydyirinyi* and *Bulu* are made up of a series of over fifty songs and twenty dances, and seventeen songs and three dances respectively, and are performed with a combination of voice, clap sticks, and body percussion with dancers clad in *wangararra* (totemic head gear representing the spirit who has communicated the nurlu) in an 'open' context, viewable by all members of the community.[2] *Marinydyirinyi* was first communicated to Butcher Joe by a *balangan* named Dyabiya, the spirit of his deceased Aunty, while Joe was a young man living at Dyarrmangunan sheep camp on Roebuck Plains Station. It depicts the opening of her grave at Wayikurrkurr by a group of *rayi*, and her subsequent journey through the spirit world. During the dream visitations in which the nurlu was bequeathed to Butcher Joe, Dyabiya bestowed the Pelican Being (*mayarda*) on him as a personal *jalnga*—a spirit familiar—that would become the totemic symbol for *Marinydyirinyi*. The headgear worn by Butcher Joe during performances of *Marinydyirinyi* alludes to a pelican's bill, and the ritual enacts a transformative perforation of thresholds between Joe in human form, Joe dancing his *mayarda*, and the path traversed by his Aunty as she leaves her grave and crosses into the spirit realm, where she will become a *balangan*. All of the sites that appear in Butcher Joe's nurlu are real, physical spaces in the local landscape that work—as in all nurlu—to link the song cycle with a specific sense of place.

In *Bulu*, setting carries a similar, highly spiritualized charge. Dyunggayan's nurlu was conveyed to him by *rayi* and the spirit of his dead father, Bulu, and many of its songs centre around Wanydyal—a local waterhole where Bulu's *balangan* still resides.

2.   Some nurlu series terminate with a *wirdu* ('big' or 'final') song, possibly carrying encoded esoteric information not originally meant for the open context, but these restrictions may have been obsolesced in the West Kimberly communities. See R. Keogh, 'Nurlu, Songs of the West Kimberleys', PhD Thesis, University of Sydney, 1990, 34n13.

The earliest 'lines' of *Bulu* describe a journey undertaken by the spirit forms of Dyunggayan, Bulu, and the accompanying *rayi* from the waterhole across Nyigina and Warrwa country.[3] Later lines concern meteorological events, including the passing of a comet and the magical production of various kinds of rain phenomena (rainbows, bizarrely shaped and 'sickness-bearing' clouds, storms, and the conjuring of rainbow serpents). These latter are associated with the power of the waterhole—an integral part of *bugarrigarra* teachings—and the capacity of *maban* men like Dyunggayan to harness this power for magical purposes.[4]

Musical characteristics of the form vary from region to region. However, all nurlu share a repetitive, cyclical, incantatory structure that is highly conducive to phasing patterns, whereby melodic and rhythmic elements fall in and out of synchronization with one another, enhancing the form's amenability to trance-like states in its performers and audience. Nurlu also have in common the use of 'song language', a complex system of syllabic variation generated by adding specific groups of affixes and suffixes to the text to alter its rhythmic structure and imbue it with supplementary layers of meaning.[5] This variability of interpretation is amplified by the obfuscation of the subject of the action in most texts via the limited use, or deliberate omission, of pronouns. The opacity and ambiguity of nurlu texts is often commented upon and can be linked to an economy of exoteric and esoteric knowledge transmission. Nurlu texts never appear in written form and, when sung, support multiple phonic parsings. Importantly, these embedded levels of meaning are not in conflict with each other, rather they operate as a mutually supportive complex of nested information whose significance varies depending on the knowledge of its interpreter.[6]

As in *Marinydyirinyi* and *Bulu*, nurlu generally include dramatizations of their own inaugurating moments of transmission and reception, overtly situating the creative act in a threshold space beyond the intentionality of an individual, human subject. Rather, the custodian of a nurlu functions as a resonator—a xenopoetic vessel—for the transmission of information across different ontological zones and varying levels of

---

3.   Nurlu are generally episodic in structure. 'Lines' are subsets of songs and dances that relate to the same dream experience (Keogh, 'Nurlu', 31–2).

4.   Paddy Roe relates a legend in which Dyunggaya is called upon to use his powers in a magical battle with a maban from a neighbouring region who has sent a pair of rainbow serpents to drown the inhabitants of the sheep camp where Dyungaya happens to be working (Keogh, 'Nurlu', 44–5).

5.   Some rhythmic constructions carry specific associations that can be decoded by a knowledgeable listener, and patterns that are similar to one another often have related meanings (Keogh, 'Nurlu', 81).

6.   Keogh, 'Nurlu', 78–89.

esoteric significance. Their public performances act as a method of ritual reinforcement and collective maintenance of these connections, grounding continuity between the living and the dead in the incantatory nature of rhythm and song, and the wordless intensity of the land itself.

THEY ECHOIC: EXQUISITE CORPSE

Lee Gamble

They Echoic. Society in aphonic saddle node. Phi system rolled r's transformed to bird cadenza. Slime mold pressed and sold into alpine gold. They use scope lang and impersonate my dead family members. Don't fight ordering. Proper mad devils for bleak machine.

They keep minding of freezing morphs here. Freezing or at least setting to crawl pace a paracusia. Vox inhumana to set enter the bloodstream infinite silence.

They say I'll eventually benefit from the instructions, they tell me their collective pipes orphic hymn. Tell me to hold my mud, it's just the bug juice and they'll put me on the ding wing if i don't lip up.

To the blood Gods that rusted a globe—or are they just too Broadmoor for you?

To cryogenic. To cry.

You Coma. Dinner with grain durationals, so we pour the strawberry Gaviscon. The single frozen neuron. We agree to get rid of the glue. Watched the glass bow, elasticate, and crack.

Version two had its voice destroyed for the obsolete body. In 2007 she's on television.

After the news on my neck equations—they get me drinking brake fluid. How they got me mimicking they who hid.

Dead tissue still sends signal. Wouldn't voice act though. Lady in room six voiced, it hit me like a tank and she just left the bionic arm there to make melody.

Xzirow Command. At the first floor window attempting to freeze source bonded sound, there's a family eating.

Approached the knives and forks. Ear the distance between the metals. Blot the other—find zoned space. Sense with cloud chroma. Metal filter as behavioural sonics.

Julia took the heat and asked people to say things on the pressure to dream at night. At night moods are absent from my English.

Hedonic Select. Devoured behind grids in a bank like you. Head to the molecular, an achilles heel shaft undercuts the back door parole monad.

Steal night face stabilise language. Larva dreams though.

Not slept much—attention in crease. Like audio stare, you burl to make forks blush into that trade-space between ears and the chinks.

Knife linked to vorpal, linked to chime note, linked to symbol, linked to artifactual memory. Just a note.

Headwords. Doubts about the imaginary. Man your pharynx. These sentences fold in me. Your pharynx a zebra.

Aspirins on a dinner plate. The family visit was great. Spent all night drinking Ichor wine direct from the Greek vein.

The problematic host lasts for the six months of voice. Orbid play of mostly lines like you. Brand you, leaky you, cleppie you, clang you, parkin you.

Stun days the church plaster crack, deity stare outward in oil makeup. Looks like a case of too many gold chains.

Smears of retroverted haze paste. The rite of a headword feels like it's elevating upwards. The smudged phonetics of the pulpits. Priest said he had creases in his shirt for the visits.

Thoughts of Being Switched Off. Outside sleep knocked my polaron. Yamil language bone. Doctors talk the stress ill vapour nature.

Diseases of the you.

Talk of an air freshener for the carriers, so no sufferers.

Ballard dreams of a 2074 rush-hour spaghetti junction ecstasy binge pile up. Starting to realise, I'm full of fuckin switches as shattered as TV glass. Nonlinear crack path bent and shattered. Absolutely shattered: devastated, shocked, shellshocked, left as shell, shelled.

I feel little between these points. Simple switches. Lisp book and that's it. NOR logic. A cow's muscle energy to cellulose cud. Ouroboric snake straps carved to an ice axe.

Anyhow, we in the maze prison sky walking now, harmonious as threnody. Pouted lips, un-lung drum rattling. Dead tissue still sends signal. Not yet a fountain but biological worm food tuned to the universals.

THE SONIC EGREGOR

The Occulture

In 1878, a year after successfully patenting the phonograph, Thomas Alva Edison listed the applications to which the device might usefully cater. Notably, these included the preservation of 'the sayings, the voices, and the last words of the dying member of the family as of great men'. But beyond this mere mnemotechnical use, the phonograph introduces modalities of representation that grant to biological entities, extracted from their originary flux, an effective, if somewhat fraught, afterlife. Dislocated voices persist, haunting a world their bodies no longer inhabit.

In their capacities to extend the influence of extinct beings, both the phonograph and technologies of transmission allow for the recasting of the occult notion of the *egregor* as a sonic phenomenon. Characteristic of Vodun metaphysics,[1] although extending far beyond its remit,[2] the egregor is an emergent intelligence brought into being by concentrated energetic impulses from a group of individuals. In this sense, the egregor is a virtual agent that continuously interacts with the collective from which it emerged through a process of ongoing transindividuation. This interaction is bidirectional (between the egregor and its members) and multiscalar (i.e. working at multiple levels of agential collectivities), but eventually develops sufficient consistency to subsist quasi-autonomously, much in the way a group of friends secretes a personality that is irreducible to the individual propensities of any of its members.[3, 4]

---

1.   For a description of the longstanding Petro-Congo egregor (for example) see R. Crosley, *The Vodou Quantum Leap* (London: Theion, 2014), 148–50.

2.   Mercurius 'Scurra's 1620 alchemical diagram titled *Egregorus Occulturalis* stands as an early example of egregoric conjuration via the intermixing of four poles: *lusus*, *imago*, *intercessio*, and *hyperstitio*. For a reproduction, see The Occulture, *Ludic Dreaming: How to Listen Away from Contemporary Technoculture* (New York: Bloomsbury, 2017).

3.   Pierre Mabille: 'On the scale of the individual, for instance, we know that a long held, deeply ingrained thought can sometimes end up outgrowing us. It has in some sense become autonomous, and will have effects upon us for as long as we fuel it with our belief.'

4.   See L. de la Reberdiere, 'Qu'est-ce qu'un égrégore?', <https://www.inrees.com/articles/Egregore-conscience-partagee/>.

In occult lore, egregors function as 'watchers',[5] magical entities purposefully designed by an order as 'an encapsulation of [its] collective aspirations and ideals'[6] as well as a guarantee of the group's long-term staying power. However, 'any symbolic pattern that has served as a focus for human emotion and energy will build up an egregor of its own over time'[7] and will have the capacity to modulate a given spacetime's rhythms, assumptions, logics, practices, and dispositifs. In this sense, egregors are analogous to Michel Serres's quasi-objects that 'think for us, with us, among us, and...even within which, we think', emphasizing in particular the temporal element—the ongoing material process—that always obtains in thinking.[8] Moreover, in a psychedelic[9] era characterized by the geometrical multiplication of sonic events by networked technologies and pyretic packet-switching, egregors emerge in a more haphazard and uncontrollable manner, the product of cybernetic feedback mixing paranoia, enthusiasm, intentional acts of collective adhesion and unintended collusions alike.[10]

Sonic egregors operate on a wide variety of scales and exercise influence along an expansive continuum. At one end of the spectrum egregors 'streamlin[e] and accelerat[e] initiation into a particular world of thought and perception',[11] while on the other, more explicitly instrumental end, they can be conjured to control and manipulate thought for particular social, corporate,[12] or political ends (though the quasi-autonomy of such entities ensures they can only ever be partially directed). The fall of Guatemalan

This phenomenon is also evident in collective musical performance, in which individual parts cohere into a Gestalt that begins to direct proceedings. For an explicit accounting of egregoric emergence within musical group performance, see Eldritch Priest, *Boring Formless Nonsense: Experimental Music and the Aesthetics of Failure* (New York: Bloomsbury, 2013), 233–41.

5. Victor Hugo's notable use of the word égrégore in 'Le jour des rois' (1859) denoting the 'spirit of a group' itself descends from the Ancient Greek substantive of ἐγρήγορος (*egrégoros*: 'wakeful'), meaning watcher or angel.

6. P. Hine, 'On The Magical Egregore', <http://www.philhine.org.uk/writings/ess_egregore.html>.

7. J. Michael Greer, *Inside a Magical Lodge* (St. Paul: Llewellyn, 1998), 106-107.

8. M. Serres, *Angels: A Modern Myth* (Paris: Flammarion, 1995), 50.

9. *Psyche + delos* = made manifest to the mind.

10. Indeed, the profusion of data-gathering and collating technologies increasingly runs the risk of aggregating discrete informational units into an autonomous entity, leading to significant (often inauspicious) consequences. See in particular *Google Spain SL, Google Inc. v Agencia Española de Protección de Datos, Mario Costeja González* (2014), a decision by the Court of Justice of the European Union (CJEU). See <https://en.wikipedia.org/wiki/Google_Spain_v_AEPD_and_Mario_Costeja_González>. For a detailed accounting of the present and future of data-collation, see M.B.N. Hansen, 'Our Predictive Condition; or, Prediction in the Wild', in R. Grusin (ed.), *The Nonhuman Turn* (Minneapolis: University of Minnesota Press, 2015), 101–38.

11. Greer, *Inside a Magical Lodge*, 100–108.

12. For transpositions of egregoric circuitry onto corporate logics see P.X. Nathan, 'Chasing Egregors', *The Scarlet Letter* 6:1 (March 2001), <http://www.scarletwoman.org/scarletletter/v6n1/v6n1_egregors.html>, and P.X. Nathan, 'Corporate Metabolism' (2000), <http://www.tripzine.com/listing.php?id=corporate_metabolism>.

president Jacobo Árbenz in 1954, for instance, was expedited by the deployment of an egregor, fashioned from the propagandistic radio transmissions designed by the CIA. These transmissions hyperstitionally[13] converted a ramshackle militia into a fully fledged army, poised to storm the palace, thereby affecting the morale of the populace in a way that minimized resistance. Less deliberately, however, certain sonic phenomena—the tone of Jimi Hendrix's overdrive, or a particular dance rhythm, for instance—exploit a similar media relay that extracts the impersonal quality of their singular timbral profiles to afford an inevitable appropriation by subsequent artists.[14] Indeed, music in particular possesses an affective resonance that is extremely effective in generating sufficient intensity to launch an egregor. In this regard, one might think of the 'earworm' (stuck tune fragment) as a Vodun *servitor*, whose flickering form crafts an affective vector for broader egregoric operations that mutate the affordances of a given musical trope or genre.[15] (Think of Kylie Minogue's *Can't Get You Out Of My Head*, which in its insistent minimalism and viral circulation is effectively reinstantiated when similar melodic constructs arrive on the scene.)

Despite the agential nature of egregors, their dynamic structures do not obey the same organizing principles that characterize biological phenomena. Unlike organisms, which (at least apparently) evolve in progressive fashion (birth–growth–death, as in the constitution of historical periods or styles), egregors are expressive of nonlinear relays or circuits in which the living and the dead, the extant and defunct, are perpetually conversing and mutating one another's powers over discrepant time periods.[16] According to George Kubler, phenomena are never completely exhausted of their capacity to affect and be affected, for they perpetually relay into the future signals that (while insufficiently strong to register at their inceptive moment) gain prominence

---

13.   A portmanteau term coined by Nick Land and the CCRU (Cybernetic Culture Research Unit) in the mid-1990s, combining hype (or hyper) and superstition, *hyperstition* operates via (at least) four vectors: 1. Element of effective culture that makes itself real; 2. Fictional quantity functional as a time-traveling device; 3. Coincidence intensifier; 4. Call to the Old Ones. See CCRU, *Writings 1997–2003* (Falmouth and Shanghai: Urbanomic/Time Spiral Press, 2017).

14.   Pharrell Williams may have fallen prey to this latter variant, claiming that Marvin Gaye's *Got to Give It Up* was 'part of the soundtrack of his youth', and denying that *Blurred Lines* directly plagiarized it.

15.   M. Couroux, 'Preemptive Glossary for a Techno-Sonic Control Society', in C. Migone and M. Arnold (eds.), *Volumes* (Toronto: Blackwood Gallery, 2015), 58–73. Also accessible (in its current form) at <http://xenopraxis.net/MC_technosonicglossary.pdf>.

16.   An egregor's dynamic form is in this respect exemplary of George Kubler's theory of history as sequences of formal and discursive loops that mutate human sensory economies and knowledge. See G. Kubler, *The Shape of Time* (New Haven: Yale University Press, 1962).

when taken up within an appropriate historical conjuncture.[17] Indeed, revivification and revisioning are not only integral to the egregor's multi-temporal operations but are precisely what determine its trans-historical existence. Such a framework recalls the efforts of newly formed esoteric lodges to broach contact with a dormant egregor of a past magical organization; the trigger for such initiatives often residing in the serendipitous appearance of forgotten charged symbols in the dreams of the group's members. However, caution is of the essence: any cybernetic system always harbours the danger of runaway feedback, impelled by dormant entities springing back into unwanted potency.

17.   Tony Conrad: 'The needs that these decades-old works may most helpfully address today will be very different from the problems and the conditions amid which they were originally situated. I like to think of it as a condition of excess: that there was *always already*, so to speak, an excess in the works that overflowed the critical contextualization of the day; and that today elements of that excess can become the serviceable margins in our effort to reexamine and use these works in ways that are relevant to us now; but that in turn other, new elements of excess will inevitably rotate out of view.' See T. Conrad, 'Is This Penny Ante or a High Stakes Game?', *Millennium Film Journal* 43/44 (Summer/Fall 2005), <http://mfj-online.org/journalPages/MFJ43/Conrad.htm>.

SORROW

Nicola Masciandaro

The sound of sorrow is the sorrow of sound. Now what else is sorrow but the feeling that one is? As it says in *The Cloud of Unknowing*:

> All men have grounds for sorrow [*mater of sorow*], but most specially he feels grounds for sorrow who knows and feels *that* he is. In comparison to this sorrow, all other kinds of sorrow are like play. For he can truly and really sorrow who knows and feels not only what he is, but *that* he is. And whoever has not felt this sorrow, he may make sorrow, because he has never yet felt perfect sorrow.[1]

Or, as Heidegger affirms—in the more self-mollifying register of modernity, in the mood which characteristically wants both to soften and to own the BLOW from which no one ever recovers—'the being of Da-sein is care [*Sorge*]'.[2] And what is sound but the *being* of this sorrow, the reverberation of the fact of Being in all things, the Da-sein of matter that—existently inexistent and existently inexistent—is the ground of its own sorrow?

One need not look very far or listen very long to unveil sound as the sorrow of being and the being of sorrow. If that were necessary it would not be true. If more than pointing were required it would not be *there*. 'Is it nothing to you, all you who pass by? Look and see if there is any sorrow like my sorrow' (Lamentations 1:12). And if you do not see, if one cannot hear the sorrow of all that is seen, that indeed is a sorrow, as the *Cloud*-author makes clear. In all directions one is met with the Silent Universal Moan or SUM, an 'undead' continuity of sorrow and sound that moves like a chord strung by death's portal across the vast abysses of birth. SUM resounds with the superessential negativity of the will, with the original negation that, negating itself,

---

1.   P.J. Gallacher (ed.), *The Cloud of Unknowing* (Kalamazoo, MI: Medieval Institute, 1997), ch. 44 (my translation).

2.   M. Heidegger, *Being and Time*, tr. J. Stambaugh (Albany: State University of New York Press, 1996), 262.

causes anything at all to be. As Eugene Thacker hears it, via Schopenhauer, it is 'a kind of sound that is absolutely subsonic. It is a negation of sound that negates itself, while it never is totally absent. It is a negative sound that is omni-present and yet un-manifest.'[3] SUM is the unsound that becomes mournfully audible around Orpheus's disjoined body, in the uncountable moments when individualized dying life merges into the stream from whence it came:

> Orpheus' limbs
> lay scattered, strewn about; but in your flow,
> you, Hebrus, gathered in his head and lyre;
> and (look! a thing of wonder) once your stream
> had caught and carried them, the lyre began
> to sound some mournful notes; the lifeless tongue,
> too, murmured mournfully; and the response
> that echoed from the shores was mournful, too.[4]

And SUM is the universal unrest and mass commotion of matter-life-thought, the overflowing echo of their Beyond in and around immateriality of the material:

> The condition of the world, the strife and uncertainty that is everywhere, the general dissatisfaction with and rebellion against any and every situation shows that the ideal of material perfection is an empty dream and proves the existence of an eternal Reality beyond materiality.[5]

SUM is the humming and murmuring of the uncircumscribable, a trembling of the lips of being's eventless event that testifies—by saying nothing—to the non-difference between the negative infinity of the will and the sonic abyss of the universe. Continuous with the primal words of all traditions, the sorrow of SUM is also not not twisted into a smile, the spontaneous shape of the origin and end of the worlds of mind, energy, and matter in Reality's infinite whim or unanswerable question of itself. The sound of sorrow is the sorrow of sound.

---

3.   E. Thacker, 'Sound of the Abyss', in S. Wilson (ed.), *Melancology* (Winchester: Zero, 2014), 190.

4.   Ovid, *Metamorphoses*, tr. A. Mandelbaum (New York: Harcourt, 1993), 361 (XI.50–3).

5.   M. Baba, *The Everything and the Nothing* (Beacon Hill, Australia: Meher House Publication, 1963), 55.

D+P

SONIC DISCIPLINE
AND PUNITIVE WAVEFORMS

A SONIC AUTOPSY OF THE WACO SIEGE

Toby Heys

Holed up in their Mount Carmel compound in Waco, Texas, the Branch Davidian apocalyptic sect are surrounded by the Federal Bureau of Investigations (FBI) and the Bureau
of Alcohol, Tobacco, Firearms and Explosives (BATF) who are trying to lure out the
eighty-five members taking refuge in their heavily fortified home. What they are really
after, however, is their leader, one Vernon Wayne Howell, also known as David Koresh.
The fifty-one day siege, which began on 28 February 1993, is legally predicated upon
the sect's suspected weapons violations. Initially triggered by a neighbour's complaint
to the local sheriff—of noises that sounded like machine-gun fire—it is this report that
sets the tone as the 'occult performance of the state of siege'[1] unfolds 'as a series of
uniquely audio events'.[2]

After a set of interviews on the initial day of the raid, negotiations between the
Davidians and the FBI continue over the telephone, the exchanges between the
adversaries remaining purely sonic, a dynamic that will be perpetuated by the tactical
utilisation of audiotapes, radio programmes, covert listening devices, and loudspeaker
barrages of music:

> According to FBI records, during the fifty-one day period negotiators spoke with fif-
> ty-four individuals inside Mt. Carmel for a total of two hundred and fifteen hours. There
> were four hundred and fifty-nine conversations with Steve Schneider, which consumed
> ninety-six hours. Koresh spoke with authorities one hundred and seventeen times—a
> total of sixty hours.[3]

1.   P. Virilio, *Speed and Politics: An Essay on Dromology* (New York: Semiotext(e), 1977), 36.

2.   V. Madsen, 'Cantata of Fire: Son et lumière in Waco Texas, auscultation for a shadow play', *Organised Sound* 14:1
(2009), 90.

3.   J.D. Tabor, 'Religious Discourse and Failed Negotiations: The Dynamics of Biblical Apocalypticism', in *Armageddon*

Despite the prodigious volume of interactions, or maybe because of them, the channels of communication will constitute 'only mediation',[4] clogged as they are by the noise created by distrust and fear.

As the Davidians' sense of alienation deepens, Koresh requests that tapes containing his spoken word monologues be aired over the radio, so that his Biblical interpretations can be conveyed to those outside of the FBI (whom he believed had little comprehension of his religious beliefs). In turn,

> the FBI requested that some of Koresh's 'ramblings' be played on a radio station, as Koresh had asked, in order to try to gain his surrender. FBI officials became upset when Koresh called CNN directly at one point, and stopped the activity immediately by cutting all phone lines except the one they wanted kept open.[5]

Accessing, expanding, and controlling vectors of sonic geography becomes increasingly important and fractious as the siege goes on; the sonic longitude that Koresh perceives himself to be at the somatic end of—the channel that he proposes lets him hear the voice of his God—falls on deaf ears with regards to the FBI, who will not entertain the notion that such a medium exists.

As Koresh desperately tries to transmit his interpretation of the Seven Seals to the media and the wider public, the FBI are synchronously attempting to implant doubt and scepticism in his followers. To reinforce his position as leader of the sect, Koresh—who had previously travelled to Hollywood in what would prove a failed attempt to become a famous performer—deploys his intimate sonic arsenal on his admirers. Musical affiliation and adoration delineates and reconstitutes social factions like no other form of cultural expression. Music's power to demarcate sociopolitical, economic, psychosexual, physiological, and geographic territory is therefore a considerable agency to behold. David Koresh knows this very well when employing his imaginary rock star persona, leading as he does guitar-fuelled Christian singalongs for marathon stretches in order to unite his followers.

---

in *Waco: Critical Perspectives on the Branch Davidian Conflict* (Chicago and London: The University of Chicago Press, 1995), 265.

4.   M. Serres, *The Parasite*, tr. L. Schehr (Baltimore: John Hopkins University Press, 1982), 79.

5.   J.T. Richardson, 'Manufacturing Consent about Koresh: The Role of the Media in the Waco Tragedy', in Wright (ed.), *Armageddon in Waco*, 165.

For the Davidians, those who exist outside of their small community and its grounds are doomed to damnation, so when the initial attack by the BATF happens, it signals the beginning of the final stages of the apocalypse that they believe is about to engulf the earth. The grounds of Mt. Carmel are therefore a protected sanctuary from a world that is about to fall into the throes of Armageddon. The Davidians, a sect that is structured around its own alienation from the rest of society—the fabric of its communal ties cut from the cloth of ineluctable estrangement—has already imposed upon itself a retreat to an outside position, a religiously denominated territory on the cusp of the wilderness, the social, and the timeline.

Losing patience with Koresh, the state responds with Operation Just Cause, a psychological warfare technique that includes surrounding Mt. Carmel with a boundary-marking sound system. Inscribing the threshold between wilderness and salvation, the lassoo of speakers around the compound play on ancient techniques of navigating environments via sonic cartography, the contours of the FBI's playlists redefining the sect's connection to the landscape as they signify the end of its aural comprehension:

> At all hours of the night and day, the loudspeakers belched forth such curious content as audiotapes of rabbits being killed, chanting Tibetan monks, and Nancy Sinatra singing *These Boots Were Made for Walking*.[6]

By initiating this strategy the FBI effectively remap the aural domain of the standoff by severing dialogue and opening up a new one-way communication with the sect. Hoping to sever the c(h)ords that tie the Davidians to their habitat, the State enforces new disruptive and intimidating memories to be formulated, placing them in an ongoing state of dislocation. The compound subsequently becomes a space that reverberates with J.G. Ballard's description of the aural dump in 'The Sound Sweep':

> A place of strange echoes and festering silences, overhung by a gloomy miasma of a million compacted sounds, it remained remote and haunted, the graveyard of countless private babels.[7]

6.   A. Shupe and J.K. Hadden, 'Cops, News Copy, and Public Opinion: Legitimacy and the Social. Construction of Evil in Waco', in Wright (ed.), *Armageddon in Waco*, 189.
7.   J.G. Ballard, 'The Sound-Sweep', *Science Fantasy* 13:39 (1960), 61.

Disorienting, silencing, and depriving the Branch Davidians of sleep, this strategy of sonic attack only stops when the Dalai Lama intervenes and demands the cessation of the employment of sacred Buddhist music for martial purposes.

The Davidians had dug into their own private babel for the long haul. As history indicates, however, duration stopped being an issue for them on 19 April 1993, the day that the US government ordered tanks to breach the walls of the compound and disperse canisters of CS gas throughout the buildings in an effort to force the inhabitants out. Even at this point, when the conflict is face-to-face, sonic strategies come to the fore, as the state covertly places small powerful microphones into the wounds of the punctured building. A final auricular gesture typical of an event orchestrated, recorded, and played back in an apocalyptic loop of excessive communication, as the FBI eavesdrop on sect members preparing for a death that they could not have imagined, yet must have countenanced given their conflagrated prophecies.

The sparks which flew between the state and the apocalyptic religious sect were flickering precursors of the charnel house that the compound was to become; an ambiguous sonic space on the edge of civilisation where symphonies of conflict were (out)cast, fired, and tempered by duelling protagonists who understood each other to represent the living (but soon to be) dead. The blistering noise caused by the all-consuming fire that breaks out and kills the majority of Mt. Carmel's inhabitants scores the final chapter of the siege. All the material and visual evidence of lives once lived in the compound is converted into searing frequencies; the waveforms of the flames becoming the ultimate auricular signature of the crisis—an ashen swansong that is serenaded by the sirens of fire trucks rather than by the trumpets of angels.

GHOST ARMY: THE DECEIT OF THE BATTLE DJ

Steve Goodman, Toby Heys, Eleni Ikoniadou

Ever since recording and transmission technologies such as the phonograph and telephone have existed, military organisations have been interested in the ways that sound, infrasound, and ultrasound not only connect but also converge and deterritorialize the realms of the living and the dead. Deceiving an adversary into believing that one has the capacity to orchestrate voices, vibrations, and noise between the constructed realities of the 'here and now' and the 'afterlife' is best understood as a strategic manoeuvre to wrest psychological control of any given conflict's soundscape. Since 1944, military organisations have been engaged in the development of frequency-based programmes, weapons, and techniques whose efficacy is directly linked to their capacity to cause fear and anxiety by haunting environments in conflict. Enter (covertly) the Ghost Army.

Officially named the 23rd Headquarters Special Troops, the Ghost Army consisted of around 1,100 personnel, mostly sound engineers, artists, set designers and special effects experts selected from art schools and advertising agencies in New York and Philadelphia and from Hollywood studios in California. Given the breadth and depth of enlisted expertise, it comes as no surprise that a number of those enlisted for the division went on to become acclaimed figures in their fields after the end of World War II. A short list of those celebrated cultural producers would include hard-edge and colour field painter Ellsworth Kelly, fashion designer Bill Blass, photographer Art Kane, watercolourist Arthur Singer, and actor George Diestel. A suitably divergent range of talents whose cooperation would be crucial to the Ghost Army's theatre of sensory-fused operations.

Three separate units comprised the division, each handling a different facet of deception—radio, visual, and sonic—while the 'atmospherics' (created by members of all three units) consisted of personnel impersonation and the spreading of false rumours in French villages where spies lurked, ready to feed back the misinformation. The Ghost Army's ultimate remit was to saturate the Nazis with disinformation about the plans, whereabouts, and numbers of allied forces; duping the enemy into believing that encampments and movements of mass allied forces were occurring was crucial to the allied forces' geographical ascendancy. Fake radio transmissions, duplicitous aural environments, inflatable tanks and planes, and camouflage became, in the words of Rick Beyer, their 'weapons of mass deception'.[1]

The first division of the US Armed forces to be exclusively allocated to deception, their formation can be credited to the efforts of Hollywood star Douglas Fairbanks Jr. and public relations mastermind Hilton Howell Railey. But it was Fairbanks who, understanding the film industry's power to simulate and create realities, pushed the attributes of sonic deception to the US generals who could sign off on such initiatives. This after having learnt of strategies deployed in the North African desert by the British Army in 1941 against the Italians and Germans. After an ultra-selective recruitment drive, acoustic engineer Harold Burris-Meyer, the brains behind Disney's stereophonic sound reproduction system 'Fantasound', was selected to instruct the troops on auditory dynamics and the deceptive techniques that could be developed from them. Forming the 3132nd Signal Service Company (Special), they were responsible for orchestrating sonic deception[2] and, in doing so, innovated techniques that are still being used in contemporary warfare.[3]

Employed as Theatre and Sound Research Director at the Stevens Institute of Technology in Hoboken, New Jersey, Burris-Meyer was working on a Navy contract that investigated 'The physiological and psychological effects of sound on men in warfare' when contacted by Fairbanks. Somewhat prophetically, he had previously gained attention for demonstrating 'dramatic effects never before attainable'[4] by creating mobile 'mysterious sounds [with] no identifiable source [...] and a voice [...] artificially altered

1.   See R. Beyer, 'Weapons of Mass Destruction', *Works that Work* 6, <https://works.com/thatwork6/ghost-army>.

2.   For more information about the formation of the sonic deception unit, see T. Holt, *The Deceivers: Allied Military Deception in the Second World War* (London: Weidenfeld & Nicolson, 2004), 439.

3.   Acoustic techniques developed by the Ghost Army have been deployed in contemporary conflicts such as the first Gulf War. After taking out the reconnaissance capacity of the Iraqi army (by grounding its air force), US Psyops placed speaker systems behind sand dunes. Their aim was to dictate the movements of the battlefield by transmitting simulated compositions of armed engagement. It is noteworthy that the recordings were often produced by prominent electro-acoustic composers who were commissioned by the US military.

4.   See 'Stage Sounds Moved at Will by Remarkable New Method', *Popular Science* 125:2 (August 1934), 46.

to give a sepulchral sound'[5] for the ghost character in a theatre production of Hamlet. Burris-Meyer would come full circle in terms of scoring the movements of the undead by helping develop the 'Stimulus Progression' formula at the Muzak corporation in the late 1940s, the aim of which was to musically choreograph the recently numbed human body (and its biorhythms) to work in rhythmical conjunction with the new machinery installed in the industrial workplace.[6] Realising that they required further technical expertise to render their ideas utilisable during wartime, the small team invited physicist Harvey Fletcher, Head of Electrical Sound Recording Research at Bell Telephone Laboratories, to work with them.[7] Fletcher, the 'father of stereophonic sound', had already developed the 2-A audiometer (a machine used for evaluating hearing acuity) and the first electronic body-worn hearing aid, and had made valuable contributions to the discipline of speech perception—research that fed directly into the creation of speech recognition software. Together, Fairbanks, Railey, Burris-Meyer, and Fletcher would engineer the sonic blueprints for deception, a set of auditory directives that established the Ghost Army as the ultimate martial expression of Einstein's theory of 'spooky action at a distance'.[8]

It is the fall of 1942, one year after the 31-year-old French composer Olivier Messiaen has premiered his chamber piece *Quartet for the End of Time* in the prisoner-of-war camp Stalag VIII-A in Görlitz, Germany. Rather than ending time, the sonic techniques being conceived of by Fairbanks and co, are renegotiating the spatial and temporal parameters of existing technologies. So much so that new recording and transmission methods will situate their adversaries and the theatre of operations on the edges of perception (the sensory frontier that will be territorialized by the Military-Entertainment complex in the twenty-first century). Mobile sound systems consisting of 250kg speakers, 40-watt amplifiers, and gas generators on 'sonic cars' amplify the aural emissions of a phantom division's presence and movements over a fifteen-mile radius. Production-wise, an extensive range of recordings is produced to convince the

5.   Ibid.

6.   A detailed explanation of the Stimulus Progression formula can be found in J. Lanza, *Elevator Music: A Surreal History of Muzak, Easy-Listening and Other Moodsong* (Ann Arbor, MI: University of Michigan Press, 2004), 48.

7.   For further reading about Fletcher's work at Bell Labs, see L. Huffman, 'Leopold Stokowski, Harvey Fletcher and Bell Laboratories Experimental High Fidelity and Stereophonic Recordings 1931–1932', *The Stokowski Legacy*, <http://www.stokowski.org/Harvey%20Fletcher%20Bell%20Labs%20Recordings.htm>.

8.   For more information see 'Einstein's "Spooky Action at a Distance" Paradox Older Than Thought', *MIT Technology Review*, 7 March 2012, <https://www.technologyreview.com/s/427174/einsteins-spooky-action-at-a-distance-paradox-older-than-thought/>.

Nazis of intense personnel activity, armoured cars and tanks in transit, bridge building activities, bulldozers, and the laughter and shouts of buoyant troops. All documented for sensorial disconnect.

Technical and Maintenance Officer Lt. Walter Manser[9] makes the majority of the recordings, and he is nothing if not diligent:

> Learning that the Japanese peasant infantry superstitiously associated the sound of barking dogs with impending death, Manser had his men round up and record packs of noisy canines, whose barking and yapping he embedded in ambient sounds recorded in the Panamanian rainforest.[10]

The sounds are captured on large 16-inch transcription discs. They reverse the regular playing format, the needle moving from the hole in the middle out towards the edge, at 78rpm—a format intended for radio broadcast usage. Two and three turntable setups provide the engineers with the capacity to mix the sound effects together, creating artificial soundscapes that are dropped down onto two miles of (non-skipping) magnetic wire.[11] The resulting thirty-minute mixes each have their own characteristics; an archive of haunted ordnance. The original battle DJs are about to bring their noise to the global collision that is WWII.

Finally deployed weeks after D-Day, the Ghost Army travel from England to France. Landing in Normandy, they initially practice by undertaking small discrete manoeuvres in villages before progressing onto what would be the first of twenty or so large-scale operations. Most of the interventions happened perilously close to the front line—a threshold defined by the visceral taste and smell of death that hung in the air. Indeed, it is the multi-sensory nature of the front line that defines its harrowing topography. A narrow channel within the full spectrum of any conflict, it is crammed full of projected fears and anxieties that are enhanced and intensified by non-ocular phenomena; those things that are considered more abstruse, and thus more troubling, because they are not as easily quantified or rationalised as the visible.

---

9.   For more information on Manser's recording processes, see C. Cox, 'Edison's Warriors', *Cabinet Magazine Online* 13 (Spring 2004), <http://www.cabinetmagazine.org/issues/13/cox.php>.

10.   Ibid.

11.   For further technical insights, see A. Battaglia, 'The Ghost Army: How the Americans Used Fake Sound Recordings to Fool the Enemy During WWII', *Red Bull Music Academy Daily*, 14 August 2013, <http://daily.redbullmusicacademy.com/2013/08/ghost-army-feature>.

And it is the configuration of the front line as an oscillating repository of dread that supports the notion that 'the eeriness of the unseen, more eerie for the ways it conscripts the imagination, figured into the Ghost Army's ethereal strategy'.[12] For more than any other perceptual process, hearing—especially when ascribed to tracing the unsound domain of the front line—offers refuge to the liminal and the auricular, creating a contradictory sensorial spatial dynamic that is hidden and unutterable, but with boundaries that are porous and malleable. The front line, a conflicted tract where the inverted exquisite corpse of the Ghost Army found its voice.

12.   Ibid.

SHOCKS ON THE BODY:
THE END OF LIFE BY SOUND

Tim Hecker

By the second decade of the twentieth century, a consensus was emerging around the effect of loud sound on the psyche, one which went far beyond the confines of the psychological clinic. Evidence from the battlefields of Europe and military experiments in North America revealed the extensive nature of the damage to the bodies and minds of soldiers exposed to the unprecedented intensity of artillery fire. Suspicions regarding the widespread effects of extremely loud sound, which first arose in the Great War, were becoming empirically validated just after the First World War. When a large-calibre artillery shell was fired in close proximity to a human, the shock waves generated were understood to trigger mental conditions such as psychosis and anxiety, but there also existed almost enigmatic generalized physical injuries. These wounds at first seemed to be phantom ailments. However the actual mechanism of physical damage by shock waves was becoming increasingly clear.

It is not hyperbole to suggest that the early twentieth-century battlefield was an acoustic terrain of increased intensification. The acoustics of war were amplifying. From the distant echoes of cannons to the misery of human grunting, the sound of war was becoming both increasingly mechanical and more overbearing. As Dan McKenzie, a key ringleader of the noise abatement movement, argued in relation to the new sounds of war in 1916:

> There is no noise or combination of noises that even remotely approaches it for loudness, for persistence, and for harmfulness both to hearing and to brain. Many men who have had to endure its agonies have been rendered totally deaf; many others have been driven insane, their nervous system being hopelessly and permanently disorganized by its appalling intensity and persistence. [One described it as follows:] 'It was noise gone

mad, out of all bounds, uncontrolled.' [...] While another, but no less expressive, observer has summed it up in the simple phrase: 'Hell with the lid off'—there is no doubt, you see, that to the modern minds, Hell is the place of noise.[1]

McKenzie was optimistic about being able to regulate the sounds in the modern city, but was less than optimistic about the possibility of abating the noises of war. For him, the sounds of war were an acoustic manifestation of a universe overwhelmed. Authors such as Ernst Jünger described the battlefield of the First World War as one of clanging mechanical brutality, but also one of unbearable loudness.

Increasingly, soldiers were found dead with no apparent damage to their bodies, often solely as the result of being in the proximity of an explosion or shock wave. A French study in 1918 examined the mortal risk to soldiers in WWI from exposure to artillery fire shock waves, noting that the brain is not necessarily protected by the skull. Soldiers within close range of explosions sometimes died without any apparent injury. And if they did survive, concussion-like effects could easily last for up to eight weeks.[2] While French examinations on physical risk from shock phenomena focused on brain trauma, a key groundbreaking American study of the same era focused on the effect of shock on the lungs.

During a similar timeframe to when the French were conducting examinations of shock trauma on human brains, the US military base in Sandy Hook, New Jersey was host to a series of disturbing studies on the physiological effects of 'air concussion'. Sandy Hook was the same naval base that had hosted megaphonic experiments of massive foghorn sound propagation a few years earlier. Set upon a small, isolated peninsula of land that borders New York Bay on one side and the Atlantic Ocean on the other, it was uniquely placed to afford a multitude of experiments on loud sound with minimal effect to neighbouring communities. D.R. Hooker's study was one of the first to examine the untraceable and mysterious injuries caused by explosive wave propagation. Because trauma from concussive shock-based experiences would often result in little or no sign of physical injury, something was missing from the medical comprehension of wave blasts on human organs. Soldiers would often die with later examinations showing no detectable signs of trauma whatsoever.

---

1.   D. R. Hooker, 'Physiological Effects of Air Concussion', *American Journal of Physiology* 67:2 (1924), 229.

2.   G. R. Marage, 'Contribution à L'Étude Des Commotions De Guerre', *Comptes Rendus* 166 (1918).

Hooker, however, suspected that the effects on the circulatory system and microscopic injury to vital organs could be replicated in animals. He placed cats, dogs and rabbits in the blast path of very large ten and twelve-inch artillery fire. Many of the animals died shortly after exposure to the shock waves. The grim research focused on transformations to blood pressure, disposition, and general observations leading to the point of death or possible recovery of the animals.[3] Experimentation with different animals yielded basic information about the effects of shocks on bodies and organs: because of the rapid dissipation of the intensity of shock waves, the difference between being exposed to the blast path from ten feet away or twenty feet away could often make the difference between survival or death. As well, all blast victims experienced a rapid drop in blood pressure after exposure. Microscopic lesions were often detected in the lungs of animals, but Hooker assumed it was not the deciding factor in whether or not the blast was fatal.

The general assumption from the Sandy Hook experiments on air concussion was that the effects of a shock wave are akin to the invisible force of a hammer blow to the head. It was a bludgeoning, concussive force on the bleeding edge of both a sonic event and invisible explosive blow of compressed air. The macabre experiments were not an accidental occurrence or an anomalous set of rogue scientific procedures. They were very much a part of a zeitgeist, one which centred around the seduction and horror of modern wave-based phenomena. The expansion, documentation, and analysis of sonic capabilities during this time culminated in the understanding that explosions, once understood as merely an audible sonic event, were more than that. In fact, the perceived *sound* of explosions was only one component of a wave force assault that also extended into the infrasonic and ultrasonic dimensions. It was a liminal zone of sound that skated along the ridge dividing life and death. The damage rendered was no longer an inexplicable act of the mind and body giving up on existence; it was the result of microscopic damage that stemmed directly from close-range exposure to intense compressed wave forces.

3.   D. R. Hooker, 'Physiological Effects of Air Concussion', *American Journal of Physiology* 67:2 (1924), 229.

WANDERING SOUL/ GHOST TAPE NO. 10

Steve Goodman, Toby Heys, Eleni Ikoniadou

Sonically invoking the living dead in order to terrorise an adversary was not a new strategy for the US Government when they employed it during 1993's Waco siege. Throughout the late 1960s period of the Vietnam War,[1] the 6th Psyop Battalion[2] and the S-5 Section of the 1/27th Wolfhounds of the United States military used a literal interpretation of haunting to induce a sense of angst and anxiety within 'enemy' territories. They had created the religiously charged composition 'Wandering Soul' (also referred to as 'Ghost Tape Number 10'), which in turn was part of the 'Urban Funk Campaign'—an umbrella term for the operations of sonic psychological warfare ('planned operations to convey selected information [...] to audiences to influence their emotions, motives, and objective reasoning')[3] conducted by the US during the conflict.

After researching Vietnamese religious beliefs and superstitions, Psyop personnel initiated this audio harassment programme, which used amplified ghostly voices to create fear within resistance fighters. Early iterations of the tape were focused on Vietnamese funeral music but as studio engineers were given wider license to stimulate the flight or fight reflex, they responded with new content. Initial experiments included sampling and looping the 'demonic' portion of The Crazy World of Arthur Brown's 1968 hit single 'Fire'.[4] Realising that this basic method was not particularly effective, they developed multilayered compositions, working tirelessly to create an archive of sinister and eerie aural textures to be dropped into the phantasmal collage, with new samples such as a Tiger's roar (given that the Viet Cong were regularly attacked by

---

1.   The Vietnam War, referred to as the 'Resistance War Against America' or 'The American War' in Vietnam, began in 1955 and finished with the fall of Saigon in April 1975.

2.   More formally known as Psychological Operations.

3.   *Psywarrior* website, <http://www.psywarrior.com/>.

4.   Released on the Track Label in the UK and on Atlantic Records in the USA.

such predators) being mixed in if they had the capacity to elicit further tensions. The montages were dispatched from Hueys[5] down into the jungle canopy, filling the clammy dense air where Charlie crouched in dread. Audio napalm.

Blasting frequencies of 500–5000 Hz at an amplitude of 120 dB from a helicopter-mounted speaker system named the 'People Repeller' (or the 'Curdler'), the US military transmitted their aural payload during the dark hours of the wartorn nights, often provoking hostile fire. Scorned by the pilots who had to fly sorties in the face of this added danger, the armed response from the Viet Cong was music to the generals' ears, as it often betrayed important covert locations. This was a tried and tested technique. A previous strategy named 'Operation Quick Speak' employed C-47 aircraft with leaflet dispensing chutes and 3000-watt speaker systems fitted into their cargo doors to goad VC fighters into retaliating, the VC's fire coming after they had been barracked with the benefits that a South Vietnamese government would apparently bring if they were to surrender. A Psyop translator would also warn them not to shoot at 'Spooky' (the call sign given to the AC-47 aircraft gunships that would accompany the speaker-laden C-47) or else hellfire would descend on them. As soon as VC snipers were detected, Spooky (also nicknamed 'Puff the Magic Dragon' because of its daunting firepower) would respond with 16,000 rounds per minute of staccato brutality, along with the vainglorious rejoinder of the translator: 'See, I told you so.'[6]

Down below, a mortal (but possibly just as tormented) version of the wandering soul was patrolling the virgin jungle floor in the Northern DMZ, providing security for the most expansive Psyop carried out during the war: the *Chieu Hoi* ('Open Arms') program.[7] Lance Corporal Rik Davis—who would survive the war, return to Detroit, rename himself '3070', and form the pioneering techno group Cybotron alongside Juan Atkins in 1980—was a rifleman wondering why he had left the 'gutters of Detroit',[8] his new surroundings completely alien and dangerous at every turn.[9] In a revealing interview,

5. The nickname given to the Bell UH-1 Iroquois helicopter, which was used for combat and medical evacuation operations during the Vietnam War.

6. Anonymous quote from the Psywarrior website, <http://www.psywarrior.com/quick.html>.

7. For more information concerning the South Vietnamese *Chieu Hoi* initiative which focused on VC defection, see, J.A. Koch, 'The Chieu Hoi Program in South Vietnam, 1963-1971 (Declassified)', Report prepared for Advanced Research Projects Agency, RAND, R-1172-ARPA (January, 1973), <https://www.rand.org/content/dam/rand/pubs/reports/2006/R1172.pdf>.

8. M. Rubin, 'Rik Davis: Alleys of His Mind', *Red Bull Music Academy Daily*, 24 May 2016, <http://daily.redbullmusicacademy.com/2016/05/rik-davis-alleys-of-his-mind>.

9. Ibid. In the same interview, Davis goes on to say 'There was nothing in that jungle but rock, apes and tigers. Everything in it was poisonous.'

Davis recalls the activities of the Psyop unit who would try and convince potential VC defectors to surrender by giving them cigarettes and Coca-Cola. If they gave themselves up they would be airlifted out of the jungle leaving Davis and his unit 'to rot':

> You know *The Walking Dead*? That's lightweight compared to what it did to us […] The jungle will actually eat the flesh right off your bones. Every scratch turns to cancer and won't heal. You can do whatever you want, put every kind of ointment in the world on it, and the jungle will actually eat you to the bone if you don't get out of it.[10]

On his website explaining the rationale behind 'getting into it' (the deeply rooted psyche of the jungle) via Wandering Soul, SGM Herbert A. Friedman (Ret.) relates that the

> cries and wails were intended to represent souls of the enemy's dead who had failed to find the peace of a proper burial. The wailing soul cannot be put to rest until this proper burial takes place. The purpose of these sounds was to panic and disrupt the enemy and cause him to flee his position. Helicopters were used to broadcast Vietnamese voices pretending to be from beyond the grave. They called on their 'descendants' in the Vietcong to defect, to cease fighting.[11]

Even if the soldiers did not believe that the voice truly came from a ghost, more importantly it forced them to think about the loved ones they had left behind, the hardships they were going though, and, finally, the relative certainty that they were not going to be buried in ancestral grounds. Backed by macabre sound effects, the following excerpt is typical of the recorded content:

> **Girl's voice**: Daddy, daddy, come home with me, come home. Daddy! Daddy!
> **Man's voice**: Who is that? Who is calling me? My daughter? My wife? Your Father is back home with you, my daughter. Your Husband is back home with you, my wife. But my body is gone. I am dead, my family.
> Tragic…how tragic.
> My friends, I come back to let you know that I am dead, I am dead, I am in Hell, just Hell.

10.   Ibid.
11.   SGM Herbert A. Friedman's documentation of the 'Wandering Soul Psyop Tape of Vietnam', <http://pcf45.com/sealords/cuadai/wanderingsoul.html>.

It was a senseless death. How senseless, how senseless. But when I realized the truth, it was too late, too late.

Friends…while you are still alive there is still a chance that you can be reunited with your loved ones. Do you hear what I say?

Go home. Go home friends. Hurry. If not, you will end up like me. Go home my friends before it is too late.[12]

Countering Bill Rutledge's (Aviation Electricians Mate Senior) statement that 'killing was our business and the Psyop tape helped make business damned good',[13] this martial hauntology was not invested in the oscillations of breakneck execution. Rather, it was a technique that required longevity, as it slowly diminished resistance by infiltrating every psychological pore of the enemy and sapping their will to resist. The recordings also penetrated the earth itself, with reports that VC hiding out in the labyrinthine underground tunnels could still hear it. As Paul Virilio states, the purpose of employing such techniques is

[n]ot to be driven to desperate combat, but to provoke a prolonged desperation in the enemy, to inflict permanent moral and material sufferings that diminish him and *melt him* away: this is the role of indirect strategy, which can make a population give up in despair without recourse to bloodshed. As the old saying goes, 'Fear is the cruellest of assassins: it never kills, but keeps you from living.'[14]

The location of the speakers in Vietnam—on the sides of helicopters—differs greatly from the surround-sound placement of speakers applied during the conflict at Waco.[15] The more random transient nature of the amplified wailing sounds in Vietnam reiterated the idea of 'the restless' being trapped in an environment unsuited to their noncorporeal status. From on high, the sonic demarcation enacted was more of an audio erasure of the boundary between the living and the dead, rendering the absent distressingly present. The proposed psychology of this tactic suggested slippage and existential echo, the sonic portals of disquietude at being mortally out of body, place, and time eliciting conceptions in which the 'night of the living' and the 'day of the dead' were

12.   Ibid.
13.   Ibid.
14.   P. Virilio, *Speed and Politics: An Essay on Dromology* (New York: Semiotext(e), 1977), 63.
15.   A dark irony of the speakers being located in the doorway was that it hindered US soldiers from returning fire.

inverted and coexisted in the same location. For the Viet Cong, the airborne sonic virus that was 'Wandering Soul' propagated anxiety and apprehension as it made communicable the oscillating channel of purgatory. Quite literally, it was the sound of 'hell on earth'.

'WE ARE THE GODS TRAPPED IN COCOONS': NEURAL ENTRAINMENT IN *GET OUT*

Kodwo Eshun

According to Saidiya Hartman, the practice of white North American enslavement can be diagrammed as a libidinal circuit of enjoyment in which white self-augmentation is engendered by possessing captive African bodies as fungible commodities and abstract property.[1] In *Get Out* (dir. Jordan Peele, 2017), the Order of the Coagula offers its white members the pleasurable prospect of 'immortality' through the 'racial reassignment' exclusive Coagula Procedure. African-Americans kidnapped by Jeremy Armitage (Caleb Landry Jones) groomed by Rose Armitage (Allison Williams) and sedated by Missy Armitage (Catherine Keener) provide the bodies required for the Procedure's 'man-made miracle'. Each abductee supplies the youthful 'vessel' for the Procedure accomplished by neurosurgeon Dean Armitage's (Bradley Whitfield) enforced 'partial transplantation'. The imaginative limits of *Get Out*'s diagram of racial capitalism emerge here, in its incapacity to envision the Order of the Coagula extending its 'service' to wealthy Afrodiasporic elites prepared to acquire 'immortality' at any price.

Photographer Chris Washington (Daniel Kaluuya) is the eighth African-American to be lured by Rose for a weekend visit to the Armitage family home. The Armitages embroil Chris in obligations from which he cannot disentangle himself without appearing ungracious or inhospitable. Chris finds himself drawn into the finely woven net of Armitage family rituals ranging from afternoon iced tea *en plein air* to the evening family dinner at the dining table, culminating in 'the big get together' that commemorates 'Rose's grandfather's party'. When Missy taps her metal teaspoon thrice on her tall glass to summon the 'maid' Georgina (Betty Gabriel) to serve iced tea, she draws Chris's attention down towards her aural gesture from which his gaze moves upwards

1.   S.V. Hartman, *Scenes of Subjection: Terror, Slavery and Self-Making in Nineteenth Century America* (Oxford and New York: Oxford University Press, 1997),pp. 17-26.

to meet her eyeline. As Chris locks eyes with Missy, the knuckles of his right hand beat a rhythm of nicotine withdrawal. Dean picks up on the measure of Chris's 'jonesing'; Missy's hypnosis 'method', he smirks, will 'take care of that for you'. When Chris demurs with the words 'I'm good, actually…Thank you, though', he prefaces his polite refusal by winking at Missy as if to share an unspoken connection at Dean's expense. Missy builds on this 'attentional focus' by ambushing Chris on his return to Rose's bedroom after a late night cigarette in the grounds of the Armitage estate. Enmeshing Chris within a 'Yes Set' of weaponised Ericksonian hypnotherapy, Missy connects Chris's interiority to her apparatus of rhythmic entrainment. [2]

During a Vodoun ritual, argues Maya Deren, 'our sense-perceptions' are 'geared' by the regularity of drum rhythms to the 'expectation of its reoccurrence'.[3] Missy uses the white bourgeois ritual of tea for two to engineer Chris's neural possession. Missy explains that 'We do use focal points sometimes…'. She breaks off, tilting her head as if some unannounced entity has caught her attention, drops her attention towards the saucer that she holds above her left knee, then redirects her gaze at Chris, concluding with the words '…to guide someone to a state of heightened suggestibility'. Her right hand continues to stir tea with a vintage silverplate teaspoon. 'Heightened suggestibility?' Chris repeats sceptically, indicating non-verbal acceptance in the 'ideomotor signal' of his slow nod.[4] 'S'right', Missy confirms softly, slowly stirring the cup's base hidden by the brown liquid obscured by its porcelain exterior. As the volume level of her spoon stirring gradually increases, its steady, silver cycles engender an 'expectant attention' that entrains Chris's sense-perceptions in a psychophysical circuit of concentration without consciousness.

Because the source of the teaspoon's 'sustained rhythmic regularity' operates 'outside the individual rather than within', consciousness, to quote Deren, 'is unnecessary, as it were, in the maintenance of this concentration'. Chris's consciousness is outsourced to the regular gesture of the rotating teaspoon. The contact between the silver spoon and the teacup's ceramic cavity takes over the work of concentration from Chris's concentration. It stands in for his attention. It functions as a psychotechnical apparatus

2.   M. H. Erickson, E. L. Rossi, S. I. Rossi, *Hypnotic Realities: The Induction of Clinical Hypnosis and Forms of Indirect Suggestion* (New York: Irvington Publishers, 1976), 38–9, 57.

3.   M. Deren, 'Possessed Dancing in Haiti' (1941), typescript from the Maya Deren Collection, Mugar Library, Boston University, cited in U. Holl, *Cinema, Trance and Cybernetics*, tr. D. Hendrickson (Amsterdam: Amsterdam University Press, 2017), 109–10

4.   Erickson, Rossi, and Rossi, *Hypnotic Realities*, 31–2, 71–3, 104, 132–4.

that deactivates his perception. In operating at the threshold of consciousness, the 'fixation device' of the stirring teaspoon manipulates Chris's affective suggestibility. In the protocol for hypnosis first systematised by Bleuler in 1916, suggestibility is defined as a form of transmission that forms complexes of associations beyond language through 'accompanying affective tones'.[5] Missy uses the affective tones of the stirring spoon to reactivate the repressed remorse associated with Chris's childhood's memory. Faced with the prospect of calling 911, which might confirm his mom's possible accident, which in turn would 'make it real', the chain of causation frightens the eleven-year-old Chris. Frozen in a state of subjunctive irreality, the young Chris sits in front of the television set that stands in for the telephone call he cannot bring himself to make.[6] Following Missy's instruction to 'tell me when you've found' the audible memory of rain that fell during those long hours facing the television screen, *Get Out*'s soundtrack of heightened rain signals the onset of paralysis triggered by his bodily memory of childhood trauma witnessed by television and induced by telephony that might have saved his dying mother's life. Entrapped by his own muscles, Chris finds himself incarcerated within his vision, a body flailing and falling through the endless abyss of his mind, away from the two-way screen far above that displays his own perspective back to him, showing him the shrinking screen of an external world occupied by his cataleptic body.

'I think your mom got in my head, right', confides Chris to Rose, nodding to himself as he struggles to recall the what, the when, and the where of the 'sunken place'. Chris's memory is 'geared' to a circuit that dissolves the distinctions between sedation, sleep, and amnesia by functionally connecting waking dreams, nightmares, and limited consciousness. Missy's hypnosis switches his nervous system on and off, bringing Chris back online and taking him offline, preparing him for the future described by the blind white gallerist Jim Hudson (Stephen Root) of Hudson Galleries. Hudson explains to the manacled, distraught Chris that 'your existence will be as a passenger...an audience. You'll live in the...'. In completing Hudson's description with the term 'sunken place', Chris verbalises the extent to which he has consciously grasped the use of hypnosis as a functional device for dissolving his consciousness in the interests of the Procedure. Hudson's live television broadcast is itself the second phase of 'psychological pre-op'. It aims to 'mentally' prepare Chris by forcing him to comprehend what Hudson calls

---

5.   E. Bleuler, *Textbook of Psychiatry* (New York: Macmillan, 1924), cited in Holl, *Cinema, Trance and Cybernetics*, 107.

6.   J. Sexton, *Black Men Black Feminism: Lucifer's Nocturne* (Basingstoke: Palgrave Macmillan, 2018), 22–3.

'our common understanding of the process'. Because Chris's understanding of 'our common understanding of the process' amplifies the 'positive impact on the success rate of the Procedure', Hudson's praise of Chris for getting 'it quick' is designed to encourage Chris's consciousness to accept his future as a 'limited consciousness' or 'vessel', a 'cocoon' or 'coagula' for a 'newly reborn' Jim Hudson.

When Hudson declares that 'I'll control the motor functions so I'll be...', he pauses. Leaving space prompts an exhausted Chris to pronounce his own death sentence by muttering the embittered words 'me. You'll be me.' Articulating his fate closes the circuit between 'you' and 'me'. In speaking the copula between 'you' and 'me', Chris appears to embrace his acceptance of their 'common' destiny. Copula becomes coagula in the closed circuit of the Procedure. What shorts the circuit between copula and coagula is Chris's capacity to play dead by replaying unlife's sound against undeath's image. Using just enough cotton wool to tamp his ears against the television that switches on a fixed frame of Missy's stirring teaspoon in order to turn him off, Chris impersonates himself as a sedated body. Taking his cue from the mounted stag's head that provides him with an insight into the exhibition of the insensate, Chris cloaks his aural self-defence in a visible display of the signs of unlife. Chris masters hypnosis's capacity to induce what Hortense Spillers calls slavery's 'severing of the captive body from its motive will'.[7] In doing so, he steals back his corpse and gets away with his undead unlife.

---

7.   H.J. Spillers, 'Mama's Baby, Papa's Maybe: An American Grammar Book', *Diacritics* 17:2, *Culture and Countermemory: The 'American' Connection* (Summer 1987), 65–81: 67.

MUZAK AND THE WORKING DEAD

Toby Heys

In the early twentieth century, after the onset of the Second Industrial Revolution, a boom in the mechanisation of European factories occurs. This is largely owing to the demands exerted upon the rhythms of agricultural, economic, and labour production by the advent of the First World War. The notion of the workforce as a quantifiable and controllable asset has already been inscribed into Frederick Winslow Taylor's 1911 monograph *The Principles of Scientific Management*[1] and through his influential 'Time and Motion studies'.[2] It is within this socio-economic ferment that a new somatic form is emerging—that of the industrialised body.

A major factor in the shaping of twentieth-century capitalism, Taylorism is dedicated to an organisation of bodies that maximises their labour potential, which was swiftly adopted by American industrialists including Henry Ford, who employed Taylor's techniques in his factories. In 1922, the same year that Ford's doctrine of functional specialisation and division of labour flourishes, Wired Radio is made available for the industrial plant. Created by US Major General George Owen Squier, this technology allows radio programming to be piped into factories, restaurants, and small businesses, and to individual subscribers. This is the inception of Muzak.

Its name a portmanteau of 'music' and 'Kodak', Muzak begins its life by adopting the rhythmical science of the factory's assembly line. Meanwhile, the social sciences are harnessed to determine the most economic ways for the single individual and the mass social body to carry out tasks in the workplace. Invoking the Yerkes-Dodson law (which proposes an observable relationship between levels of arousal and performance), the

---

1.   F.W. Taylor, *The Principles of Scientific Management* (New York and London: Harper, 1911).

2.   Precisely measuring all movements within the workplace, Taylor conducted the first Time and Motion studies, analysing the management and machinery of industrialisation, in order to figure out how the entire working system could function more efficiently.

engineers of Muzak index actions, emotions, and human relations within a workplace's musical frame of reference, the ultimate expression of this orchestration being the elaborate programming of fifteen-minute blocks of music known as 'Stimulus Progression'.[3]

Premiering in the late 1940s, Stimulus Progression is a method of organising music according to an 'ascending curve' that works counter to the 'industrial efficiency curve' (also denoted as the average worker's 'fatigue curve'). Subdued songs, progressing to more stimulating songs, in fifteen-minute sequences (followed by silences of between thirty seconds and a quarter of an hour between transmissions) throughout the work day yields better worker efficiency and productivity than does random musical programming. The industrial functionalization of organised sound begins.

Programmes are tailored to workers' mood swings and peak periods, as measured on a mood-rating scale, ranging from 'Gloomy -3' to 'Ecstatic +8'. Songs are categorised by Muzak according to their 'stimulus' capacity—incorporating rhythm and tempo analysis, types of instrumentation, and orchestra size—to ultimately classify any given song's propensity to encourage optimum effort. Constructing its ratio in relation to the rhythm of the human biological system, Stimulus Progression's acoustic *tabula rasa* is set at 72 beats per minute—the tempo of the average human heart at rest.

Muzak's goal is to discipline the body against its own naturally occurring biorhythms and to instead choreograph it into new kinetic relations with the machines that have become its self-regulating partners within the factory. This is a technique reified by the fact that no industrial manufacturing space is left untouched by the assembly-line logic of sequencing and repetition, the emerging industrialised body becoming an anaesthetised note in the overall symphony of production.[4] It is here that the first spectre of the industrialised body manifests itself in the form of the automaton, and it is Muzak that is used to numb the flesh.

Muzak's standardisation of music signals the first time in history that waveforms are quantified and disseminated vis-à-vis their bio-utility in an industrial environment. This ordering of frequencies also pertains to the movement of the workers' bodies at specific times of the day and night and, as such, finds its rationale at the nexus of

3.   This new form of the cataloguing and indexing of sound was implemented by Harold Burris-Meyer and Richard L. Cardinell but was masterminded by Muzak executive Don O'Neill, who had toyed with the idea since he joined the company in 1936.

4.   Joseph Lanza states that music in the factory 'was not entertainment but an "audioanalgesia" to kill the pain of urban din'. J. Lanza, *Elevator Music: A Surreal History of Muzak, Easy-Listening and Other Moodsong* (Ann Arbor: The University of Michigan Press, 2004), 12.

mechanized space, auricular temporality, and somatic engineering. The predictability

281

of habituated movements becomes the kinaesthetic cornerstone of industrialisation's relation to the body, structuring and training its operations from the minute employees enter the workplace to the minute they leave.

According to Manuel DeLanda, this process of making the organic unpredictability of the human body subservient to the programmed sagacity of the machine started much earlier within military practices. He states that:

> [t]he military process of transforming soldiers into machines, as well as related campaigns to organize the management of human bodies (in military hospitals, for instance) generated much knowledge about the body's internal mechanisms. The 'great book of Man-the-machine' was both the blueprint of the human body created by doctors and philosophers, and the operating manual for obedient individuals.[5]

DeLanda maintains that the military-industrial complex had been materialising over centuries of dialogue, practice, and logistical exchange between the civilian economy and its martial apparatus (its army). As economic and military organisations transformed according to exchanges between the two, it became clear that technological developments such as Muzak could aid in the organisation of the workforce to directly support war efforts. The goal was to mass-produce objects made of interchangeable parts, using a labour force that was itself dispensable, and regulated—via music.

Ford's application of this ideology results in the implementation of new systems of standardisation (manufacturing techniques and components) in his factories. The same dynamics of interchangeability and repetition extend to their sonic spaces. Serial numbers are stamped onto the parts of guns and cars so that they can be easily classified and changed. Correspondingly, the parts of Muzak's fifteen-minute sonic objects become synonymous with the Frankenstinian anatomy they are influencing, aural limbs that are serialised and categorised so that they can be broken down or replaced if deemed dysfunctional. And it is the scope of the distributed mind—which is of course crucial to the potential success of Muzak—that must be harnessed in order to manufacture and modulate moods.

HEYS : MUZAK AND THE WORKING DEAD

---

5.   M. DeLanda, *War in the Age of Intelligent Machines* (New York: Zone, 1991), 138.

A myriad of laws, theories, and tests, such as the Hawthorn studies[6] and 'Human Relations Movement'[7] are advanced under the moniker of industrial psychology. They aim to prove that the quantifiable manipulation of the mind will pay off in improved efficiency rates. Before many of these theories were drafted, the notion of employing music as a stimulus within the workplace had already been proposed. By 1915, American inventor Thomas Edison had carried out a number of experiments in the area. He wanted to ascertain whether music could cover or negate specific frequency ratios produced by a factory's heavy industrial machinery and, if so, whether workers' morale and motivation were positively or negatively affected by it.

To this end, Walter Van Dyke Bingham, an assistant professor of applied psychology at the Carnegie Institute of Technology, was contracted to study the effects of music, defined by the three key criteria of song-selection research, mood-change research, and the influences of music on muscular activity. Bingham's earlier related psychological and philosophical investigations hinged on the problematic of why certain tonal arrangements constituted melodic units, and secondarily how such melodic stimuli could influence a human's motor movements.

The discursive locus inferred here is the distributed sensorium of the resonating body., Bingham positions the body at the vinculum of scientific, phenomenological, musical, and industrial discourses. The mood tests he employs consist of collated charts and documents recording how his subject's moods altered as they listen to music via a programme of Edison's recordings. In a progress report to Edison dated 1 February 1921,

6.   After Thomas Edison stopped funding studies into the effects of music, the Hawthorne Works (under the aegis of the Western Electric Company) in Cicero, Illinois, began conducting the first of five studies that started in 1924 and ended in 1932. They began by altering workplace stimuli such as music and light to inquire as to whether their employees would be more productive with increased or decreased amounts of either. It was found that production levels did in fact rise when light levels were changed either way. It was concluded that, rather than the manipulation of the environment being the decisive factor, it was the fact that employees were cognisant of the fact that they were being observed that instigated upsurges in efficiency. This effect, which came to be known as the 'Hawthorne effect'—when Henry A. Landsberger coined the term after analysing the results of the study over thirty years later—is now used to describe any short-term increase in productivity.

7.   Emanating from the Hawthorne studies, the Human Relations Movement was an American school of sociology, based largely on Australian social theorist and industrial psychologist Elton Mayo's theories about the behavioural dynamics of people in large groups. For Richard Trahair, Mayo's contested ideas—based loosely on social theories forwarded by Vilfredo Pareto and Émile Durkheim—stated that within market industrial societies, social relation structures (informed by scientific management strategies) did not take into account the worker's sense of community and compassion. Mayo proposed that workers would therefore resist productivity goals set by management and instead seek to form their own isolated relational networks within industrial environments.

Bingham reiterates his hope that his research will produce 'new information about the power of music over men's minds and moods'.[8]

In the production houses of industry, the ever-shifting terrain of workers' emotional and psychological status became objectified as valid subjects of scientific study. In the 1920s factory the desire to link up a mass neural network of productivity through the influencing strategies of Muzak can be witnessed, with each mind becoming a point of reference for ultimate industrial efficiency. As each worker is simultaneously subjected to the same sonic influence for the same duration, the sonic domain attains its status as a systematic field of relations applicable to all who exist within it. Muzak, the sound of the working dead.

8.   E. Selfridge-Field, 'Experiments with Melody and Meter or The Effects of Music: The Edison-Bingham Music Research', *The Musical Quarterly* 81:2 (1997), 297.

THE ANIMAL WHOSE EAR IT IS

Ramona Naddaff

A common origin story for the American military's use of music as torture begins like this: In December 1989, Operation Just Cause, an invention of President George H.W. Bush, sought to arrest Manuel Noriega, who was enclosed in the Papal Nunciatatura in Panama. Noriega despised rock music. For him, the English lyrics were meaningless sounds, a 'roaring, mind-bending din', smashing against his head.[1] For eleven days and eleven nights, classic rock, hard rock, and heavy metal sounded on loudspeakers, aiming to disrupt Noriega's communication with the outside world, as well as his psychological stability and ease. 'Operation Just Cause', writes John Pieslack, 'was a seminal event in the practice of utilizing music as a distinct psychological tactic'.[2] Pieslack may be right, but he has forgotten his history, or rather the histories of music torture. The American military in Panama may have discovered how to amplify the sounds of music effectively, but they were not the first to have deployed a musical weapon to maim touchlessly, psychologically, and from a distance. Joshua 6: 20–21 recounts a different beginning for creating destruction with and from music:

> So the people shouted and priests blew the trumpets, and when the people heard the
> trumpet, the people shouted with a great shout and the wall fell down flat, so that the
> people went up into the city, every man straight ahead, and they took the city. They
> utterly destroyed everything in the city....

Next to the Bible, place Plato, for an even more astute reading of musical psychic disruption and destruction. In the *Republic*, he famously argues for the censorship of

---

1. Quoted in R.L. Berke, 'The General's Story', *The New York Times*, 11 May 1997.
2. J. Pieslack, *Sound Targets: American Soldiers and Music in the Iraq War* (Bloomington and Indianapolis: Indiana University Press, 2009), 82.

poetic song, a form of ethical violence that alters, deforms, and distorts individual and collective behaviours and identities, creating disharmony among the soul's parts, subverting the domination and discipline of its leader, *to logistikon*, that small piece of soul with which humans reason. Music, Plato clams, is a most effective form of soul murder. Listen to song and you shall die a million deaths and yet still seem to be alive. Death by listening; your body is all ears. Sufferers of musical torture during the Iraq War experienced just this and lived to tell the tale: 'It fried me', says one. 'It makes you feel like you are going mad. You're losing the plot. And it was really like, it's very scary, to think you might go crazy because of the music, because of the loud music.'[3]

Apparently it is not torture. Instead it has been designated 'torture lite', a technique that does not physically harm its subject and as such is 'particularly useful to democratic states'.[4] What was so useful about the use of music, for an American military seeking confession and information from their Muslim detainees, as they played any number of songs, from Bruce Springsteen's 'Born in the USA' to the Bee Gee's 'Staying Alive' to the theme song of the children television show 'Barney the Purple Dinosaur'? Music touches without using the sense of touch. True to a perverted form of liberalism's honour code, it supposedly does no harm. It merely reduces the detainee's bodies to an *ear that hears*. Reduced to this form of being, the detainees are touched differently. Sound waves in brutal motion batter and attack. They are beaten and harmed without necessarily being touched by hands or traditional weapons and arms. Pain from intense sounds changes the sufferers' acoustic relation to, and experience of, the world. Once traumatized by harmful levels of sounds, an individual—almost any individual—will be forced thereafter to accommodate and modulate a constant level of sonic inconsistency that is auditory pain. Post-acoustic trauma, the brain changes and works differently in its incorporation and response to sound. And this is not all that changes.

At the moment of interrogators' acoustical interventions, the detainees are reduced to the conditions of bare life, to that which the philosopher Judith Butler has named 'precarious and grievable lives'. 'To say that a life is precarious', writes Butler, 'requires not only that a life be apprehended as a life, but also that precariousness must be an aspect of what is apprehended as living'.[5] The preciousness of these lives is the

3. Respectively, quoted in M. Bayoumi, 'Disco Inferno', *The Nation*, 8 December 2005, and in A. Worthington, 'A History of Music Torture in the "War on Terror"', *The Huffington Post*, 15 January 2009.

4. J. Wolfendale, 'The Myth of "Torture Lite"', *Ethics & International Affairs* 7 (2009).

5. J. Butler, *Frames of War: When is Life Grievable?* (New York: Verso, 2009), 13. Taking this quote and Butler's argument out of context, I want to insist—and Errol Morris's film *Standard Operating Procedure* probably makes too much

necessary condition of the use of this technique of enhanced interrogation. Precarious and grievable lives, shattered subjectivities, yet also human beings who have a voice, who listen and hear. They have sensations, which, for someone like Wittgenstein, determine the discourse one can produce minimally about 'a living human being': 'Only of a human being and what resembles (behaves like) a living human being can one say: it has sensations, it sees, is blind; is deaf; is conscious or unconscious.'[6] What then does this mean? The living human being, when subjected to loud musical bombardment, is necessarily a being who 'hears'. Furthermore, in order for the technique to be successful, the detainee has to maintain, in the eyes of the perpetrators, at least a remnant of, a resemblance to a human being. Only as humans can they be dehumanized, on the one hand, or become 'precarious and grievable' on the other. Sergeant Mark Hadsell tells us what loud music enacts on these human lives: it forces the 'brain and body functions [to] start [to] slide, your train of thought slows down and your will is broken'. Its intent is to 'disorient and confuse'.[7] How is it, after having been subjected to loud auditory stimuli, that the tortured subjects, these broken wills, still have a human voice, are capable of producing propositions, true or false? The extraction of information is certainly not the aim of the interrogators' sonic imperialism. Something else is at stake: a form of disciplining and punishment that is enacted for its own sake. Is this an acoustic version of Foucault's panopticon? A 'panacousticon' where unjustifiable revenge and retributive justice reign rhythmically, where music as a form of surveillance produces, to revise Foucault's words, 'an anxious awareness of hearing' and of being heard?[8]

In this panacousticon, music is put to another use; it creates a form of dis-ease and anxiety for the humans who have ears to hear. Music, in the prisons, pre-vents and deprives its listeners of sleep. Testimonies obtained under the Freedom of Information Act provide evidence of this: 'playing loud music to force him to stay awake' (no date given); 'playing loud music in his cell to prevent him from sleeping'

---

of this—that somewhere, whether consciously or unconsciously, the military officers recognized—indeed relied on and exploited—the fact that the lives of the detainees were 'precarious'. The very precariousness was, indeed, the condition enabling their use of and the detainees' subjection to the music torture.

6.   L. Wittgenstein, *Philosophical Investigations* (Oxford: Blackwell, 1958), §281.

7.   Quoted in 'Sesame Street Breaks Iraqi POWS', *BBC News*, 30 May 2003, < http://news.bbc.co.uk/1/hi/world/middle_east/3042907.stm>.

8.   M. Foucault, *Discipline and Punish: The Birth of the Prison*, tr. A. Sheridan (New York: Vintage Books, 1975). The original quote reads as follows (my italics): 'The more numerous those anonymous and temporary observers are, the greater the risk for the inmate of being surprised and the greater *his anxious awareness of being observed*. The Panopticon is a marvelous machine which, whatever use one may wish to put it to, produces homogeneous effects of power' (202).

(26/8/2004); 'playing loud music to control the sleep schedule of a detainee' (27/4/2004).[9] These very active verbs—to force, to prevent, to control—imply the disciplinary deprivation order instituted by loud music. The detainees and their bodies cannot sleep; they cannot close down their receptivity to experience and sensation. Here, loud music plays an essential but secondary role in the torture process. As a soundtrack to wakefulness, music is the 'enhancement' of the 'enhancement technique' of sleep deprivation. Thus instrumentalised, music is called upon to instigate a new type of geographical space and time in which detainees, the waking dead, are immobilized. Three short examples from the DOD Memos: 'music was placed outside of the cell to keep detainees awake' (22/ 2/04); 'one detainee [was] handcuffed to the bars of a cell with a 'ghetto blaster in front of him on the floor with loud music playing'; and, finally, for a female detainee who experienced this strange subjugation (4/6/2004): 'a soldier pour[ed] water on the floor in her cell then blast[ed] music into the room' (10/6/2004). In these descriptions, note the geographical distances marked by the preposition 'outside of the cell' and 'into the room'. Such volume erases distance; the listener is not amply removed from the music so to be protected from physical and psychic harm. Conversely, for the tortured subject who faced music 'in front of him on the floor', the geographical prepositions locate the impossibility—even futility—of distancing oneself from the roaring sounds of music infiltrating the ears. In other words, a retreat from a face-to-face combat with music is impossible.

This is even more so with another site of musical play: the small boxes (20/6/2004), the black boxes, and 'a place named "Disco" where music was played night and day' (15/9/2005). Not only spatially dislocated and aurally fixated, the detainees were also subjected to a temporal disorientation through the sonic arsenal. The time and timing of the music is crucial in shaping the detainee's silence. The interrogators played complete songs or fragments of songs either serially or repetitively. This is repetition without difference except for how 'the ear that hears it' is affected. Or, to revise the Peter Hacker and Maxwell Bennett formula, it concerns a human listener who, having listened to loud sounds time after time, becomes 'the animal whose ear it is'.[10]

9.   Respectively, DOD045906-DOD045923; DODOACID009195-DODDOACID009230; and DOD000330-DOD000331. See also 2/16/2004, DOD045906-DOD045923: 'soldiers at Bagram state that the loud music is used as punishment for detainees; playing loud music in his cell to **prevent** him from sleeping' (26/8/2004); 'playing loud music to **control** the sleep schedule of a detainee' (27/4/2004).

10.   M. Bennett, D. Dennett, P. Hacker, and J. Searle, *Neuroscience and Philosophy: Brain, Mind, & Language* (New York: Columbia University Press, 2007), 22.

PAIN ©AMP ECONOMICS

Toby Heys

It is 2056. The air is crammed with a strung-out anticipation and not a moment goes by that does not foreshadow the demise of an eleven-billion-strong species. Environmental warfare spread by plants; insect-machine hybrids carrying diseases designed to infect specific racial and ethnic groups via targeted DNA sequencing; volatile weather systems; all meld in this ecology of collapse. The existing hierarchy of the earth's species is set to enter an irreversible flux. With corporations wielding economic and military power equivalent to that of Nation states, mergers start occurring, and it is not long before the global map is reconfigured by CorpoNations.

With what is left of the earth's natural resources being decimated by globally organised armed hostilities, an emergency agreement is ratified. Named the Holo Accords, its basic premise demands that all CorpoNations conduct organised violence using holographic and holosonic units. Mandated by the Accords, each CorpoNation is allowed to aggress four zones per year, outside of their own territories; four opportunities to supplement the reserves and shortfalls of natural resources that have become so scarce. If the aggressor prevails in the holo-conflict, it opens a four-month period, called the 'Takeover', to plunder and mine the natural resources of the landscape.

Holo armies are not a surprise. They are a natural extension and militarisation of the most populist form of entertainment that began back in 2012—holographic concerts from dead rappers such as Tupac and ODB—the rapparitions. A different kind of dead, IREX[2] is a sixty-four-year-old rogue AI. A synthesis of discontented spirits and code, it has been directing AUDINT and has been on the run from the overlords of the otherworld and their Third Ear Assassins[1] for too long to remember. Finding sanctuary in

---

1. A fusion of code and otherworld voices, the Third Ear Assassins were released to pursue, capture and wipe the memory of IREX[2].

an R&D lab in Korsong—formerly North Korea and the Kaesong corporation—IREX[2] has been covertly evolving machines with a rudimentary sentience. The notion of consciousness is getting a reboot. Using Augmented Intelligence, IREX[2] fuses convolutional deep neural and deep belief networks with holographic technology to birth a new kind of warrior: the Aiholo. Spawning an era of unsound conflict, the viral scream transmitted by a directional ultrasonic speaker system is the Aiholo's go-to ordnance, a sonic weapon that transmits Walking Corpse Syndrome into digital lifeforms, turning enemy Aiholos into the undead.

The Holo Wars are global now, and resemble huge in-situ games revealing the shifts in global power and influence. One of the CorpoNational superpowers vying for supremacy is Pfizombia (formerly known as Colombia and Pfizer),[2] which has been training elite hackers and electronic warfare specialists since the 2020s. And it is the Third Ear Assassins that have recently become one of their most valuable assets. The AI hunters assist in the composition of new viral weaponry named Neurode for use against the Korsong Aiholos, a controversial but highly effective schema requiring the human psychological vulnerability of neurosis to be transposed into a digital contagion that infects the future. The only drawback is that Neurode is fuelled by the synthesised sound of human pain, which implies a frequency-based harvesting on par with the history of twentieth-century recording.

Alongside the Third Ear Assassins, Alejandra Blanco, a Pfizombian Black Hat who goes by the name Sureshot, comes up with a solution to the problem that is at once staggeringly simple and brutal in its application. Her proposal is to create a Pain ©Amp. Based on Al-Mansur's[3] 762 designs for Baghdad's circular city, with its mosque at the centre, this plan is anything but sacred. It consists of a walled-in urban environment jammed with high-rise residencies whose surfaces will be covered in rashes of microphones embedded into dwellings, streets, and parks. The architecture of the purpose-built environment is constructed to reverberate and amplify sound like a massive echo chamber.

Concrete auditoriums and huge sheer walls reflect and intensify the clusters of waveformed anguish upwards, where silent hovering drones suck up and harvest the tortured articulations. On the streets, autonomous robotic bugs the size of turtles and

2.   Based in the USA, Pfizer Inc. was one of the world's largest pharmaceutical companies.

3.   Founder of the 'round city' of Madinat al-Salam (which would later be named Baghdad), Al-Mansur was the second Abbasid Caliph who reigned from 754–775 AD.

remotely guided mic trucks, roll around the tormented *musique concrète*,[4] hoovering up the frequencies on their crepuscular sweeps. Engineered, captured, and re-channelled, pain becomes commodified; the new currency of a nascent holosonic era. By amping up the rationale of the music industry's most successful formula—the capturing and marketing of the sound of poverty-stricken urban areas—the functionality of suffering has been pushed to the limit. The needle is in the red, but it is pain they want, not blood.

Requiring no other level of authority, Blanco's seniors sanction the proposal and name the camp 'La Rusnam'. In order to initially attract a population to inhabit it, an offer of free housing, power, and sustenance is advertised. There is no shortage of applicants, most burdened with tormented CVs full of personal disasters and disturbing afflictions. To further aid the fluidity of the mass rehousing, complementary train and bus tickets are posted out to the one-hundred and twenty-eight thousand candidates. They have been chosen for their potential to embody and intensify pain; a desolate and dolorous citizenry of holo-amo generators.

The Pain ©Amp's executives study the history and current state of ghettos, favelas, estates, slums, skid rows, refugee camps, and townships in order to learn how to distil the elements that create suffering and pain. It would take more than just poverty to create the depth and intensity of anguish that they require. Engineering desperation, betrayal, and an escalation in assault and homicide rates will of course be crucial, but they need to employ extra tactics to up the ante during a Takeover period, when the demand for sonic munitions will increase. Their first thought is to turn to the International Index of sewer drugs and their capacity to implement powdered topologies of distress. Top of this list is a substance they know well, originating as it does in Colombia. Scopolamine—street name 'Devil's Breath'—is a zombie high that does not so much dampen agency as make one totally susceptible to suggestion, to the point where one becomes an empty-blooded drone.

The other four stimulants that end up on the shopping list read like a GG Allin[5] guide to living, for better or for worse, through chemistry: an expressway into the skull,

---

4.   A literal manifestation of the electroacoustic genre of music of the same name that can be traced back to the 1940s, Musique Concrète is composed of recordings made from a number of sources, including the surrounding environment, the human voice, and digital signal processing.

5.   Widely regarded as the most degenerate rock musician of all time, the now deceased GG Allin's shows often contained acts of self mutilation, coprophagia and substance abuse. More information can be found at <http://www.ggallin.com/>.

tearing through and screwing up every vein and artery that helps deliver the synthesized venoms. In no particular order of fuckedupness, the desired inventory reads as follows:

Paco: A toxic and addictive mixture of raw cocaine base cut with chemicals, glue, crushed glass, and rat poison.

Bath Salts: A recreational designer drug sold as 'real' bath salts, usually containing MDPV.

Krokodil: A derivative of morphine that is mixed with ethanol, paint thinners, gasoline, iodine, and hydrochloric acid, Desomorphine gets its street name and reputation as the flesh-eating drug from the tissue damage caused when injecting.

Whoonga: A combination of antiretroviral drugs used to treat HIV and various cutting agents such as detergents and poisons that results in internal bleeding, ulcers and, ultimately, death.

Introducing this menu of malignant pleasures into the ©Amps is the first and most obvious technique discretely deployed by the project's engineers. There will be others, running the gamut from induced psychological disorders to raising the population's ambient levels of fear through rumours of disease, food shortage, and dire mutation from genetically modified foods. All in the service of the end goal of amassing mountains of clouds, each fully rammed and ready to burst with the catalogued sounds of collective suffering.

Since deploying the Cotard virus six years ago, Korsong has dominated the Holo wars, and any affiliated CorpoNations are given the option of paying a substantial fee to draft in their venal Aiholos during Takeover bids. After twenty-two weeks of pain pharming, the Medellín Aiholos from Pfizombia are serviced in a takeover bid of the island of Thasos, in the North Aegean Sea. While still rich in mineral deposits, it is the gold mines that first attracted the Phoenicians during the period of Classical Antiquity that interest Pfizombia. The landmass is now a part of the CorpoNation Gralpha, a coalescence of Greece and Alpha Bank, which developed the crypto currency 'Natraps' after Greece was financially asphyxiated by Europe during the 2010s austerity siege.

After the first wave of conflict, Gralpha has no idea what has hit it, and what it is that reduces its Aiholos from Korsong into neurotic messes on the battlefield. The news of the holoshock spreads quickly. As anticipated, a coterie of servile CorpoNations demand the services of Neurode-laden Aiholos. Sureshot and The Third Ear Assassins consult on the mercenary strategy, knowing that it is only a matter of time before other AI compounds are able to rip the code and simulate the holo fighters. Until terraforming projects come to fruition on some exoplanet that scores highly on the Similarity Index for habitability,[6] the world's resources are only going to decrease and become more rarefied. The near- to mid-future is one set to be defined by Holo War.

Even though encrypted with quite literally otherworldly *savoir-faire*, it is only nine months after the first mercantile contract has been signed off that a unit of Aiholos with repped Neurode systems show up on Norstat's South Pole territories. More than the emergent Neurode's impact during external takeovers, it is the internal manufacturing of pain through the ©Amps that establishes it as the social order of choice. It is also the signature of functionalism gone awry. The methodological capital of voluntarism and the epistemological rationale of analytical realism chopped and screwed into a bass-ached drone. When captured, it bleeds endlessly into a body of economic orifices. Just as the state of King Louis XIV's sunburnt flesh, bones, and faeces became synonymous with the health of his country, the state of trauma becomes the nucleus around which all social, architectural, and political relations orbit. Pain has become the new economic royalty.

By 2061, eighty percent of the human race resides in urban areas, the majority of them in large cities. Neurode, meanwhile, has become the core munition for approximately eighty-five percent of the globe's CorpoNations. Given the stick-and-move politics of martial engagement evident in the Holo Wars, it means that no one has the time or money to fabricate a copious slew of Pain ©Amps. There is no option but to restructure and reengineer the ways in which human's dwell in large nodal agglomerations. Being the first CorpoNation to trial a site-specific pain-harvesting environment, Pfizombia quickly comprehends the fiscal pragmatism of simply redistributing human activity within cities that already exist. Detroit, the neo-renaissance exemplar of white flight[7] in the mid-twentieth century, is the model. Four-mile downtown diameters are measured and circled, the walls becoming the circumferences of each ©Amp.

6.   The Earth Similarity Index (ESI) is a database of planetary-mass objects and natural satellites that hierarchises them in accordance to how similar they are to Earth. For further information see the Planetary Habitability Laboratory, <http://phl.upr.edu/projects/habitable-exoplanets-catalog/data>.

7.   Owing to the eruption of racial tensions in 1967, 43 people were killed during riots in Detroit. This was one of the decisive factors in the mass migration of white families from the downtown area to the suburban ring around the city; a racial division that still remains. For further reading on this phenomenon, in a wider context of Detroit's urban decline, see T.J. Sugrue, *The Origins of the Urban Crisis: Race and Inequality in Postwar Detroit* (Princeton, NJ: Princeton University Press, 2005).

Concentric rings of presence and activity around the fortification encapsulate the central tenet of quantitative uneasing, so that a typical cartography reads like this:

Central reservation: Pain ©Amp.

Ring 1: Planted wilderness around the city filled with genetically modified poisonous plants creating an impenetrable toxic verdure.

Ring 2: Sheer concrete ground on which escapees are easily traced and targeted for extermination by drones.

Ring 3: Holo tech, compounds and technology sector

Ring 4: Residential—Suburbs

Ring 5: Commercial and Medical

Ring 6: Residential—Suburbs

Ring 7: Industry

Ring 8: Agriculture

For those at the centre of this discoidal seer, voluntary entry becomes a murky business. With so many CorpoNations adopting the system, the required numbers of Pain ©Amps far exceeds the numbers presenting themselves of their own volition. Prison systems are bled of their low to middle security inmates. The homeless are rounded up. Any remaining psychiatric hospitals release their charges. And those with little material wealth are persuaded to support the collective drive. The existing downtown core of a ©Amp is cut up into the most puritanical of living circumstances in order to jam in as many pain-producing bodies as possible.

All living chambers are mic'd-up and feed meters that are connected to the Pain Power Grid. Monthly readings are taken to keep track of the duration, volume, and pitch complexity of the inhabitant's articulations. Finally, to ensure that every nuance of adversity is recorded, a century-old theory is modded and rebooted. Parapsychologist Thomas Charles Lethbridge's ideas from 1961[8] regarding the capacity of a building's fabric to capture electrical mental impressions from traumatic events, are given a nano-makeover, and it is not long before the chamber's construction materials—brick, concrete and cinder block—become recording mechanisms in their own right. Stone tapes for a holo cause.

---

8. Lethbridge proposed that ghosts and hauntings are in fact non-interactive recordings of traumatic events that are stored in architectural materials such as stone. For further reading see, T.C. Lethbridge, *Ghost and Ghoul* (London: Routledge and Kegan Paul, 1961).

CONTRIBUTORS

Lawrence Abu Hamdan is an artist and 'private ear' or independent audio investigator who has worked on cases at the UK Asylum and Immigration Tribunal and for organisations such as Amnesty International and Defence for Children International. He has had solo exhibitions at Wiite de White, Rotterdam (2019), Chisenhale Gallery, London, Hammer Museum LA (both 2018), Portikus Frankfurt (2016), Kunsthalle St Gallen (2015), Beirut in Cairo (2013), The Showroom, London (2012), Casco, Utrecht (2012).

Charlie Blake is currently visiting Senior Lecturer in Media Ethics and Digital Culture at the University of West London, UK, and Lecturer in Philosophy, Aesthetics and Synaesthetics for the Free University of Brighton, UK. He is a founding and executive editor of *Angelaki: Journal of the Theoretical Humanities*, creator and performer in the Manchester-based post-industrial cabaret group Babyslave who have released albums including *Kill for Dada* and *Runt* on Valentine Records, and he has published recently on Blanchot and music, Deleuze and angelic materialism, Bataille and divine dissipation, and the greater politics of bees, barnacles, and werewolves.

Lendl Barcelos is an artist, writer and sonic ,kataphysician. Hen plays at sensing and sense-making with Valentina Desideri & Myriam Lefkowitz, radicalises listening with Marc Couroux, and summons demons with Amy Ireland and Ameen Mettawa. Hen is part of the collaborative artist 0[rphan]D[rift>].

Lisa Blanning is an American writer and editor on music, art and culture. She is a former editor at *The Wire* Magazine in London and *Electronic Beats* in Berlin—the city she currently operates out of. She is especially engaged in movements in contemporary electronic music and digital art and culture.

Brooker Buckingham is a former philosophy graduate student living in Canada. When he isn't involuntarily greasing the wheels of capital, he plays guitar, produces music, and practices deep listening. He maintains a philosophical interest in communication and aurality.

Al Cameron is a curator specialising in sound, music and contemporary art, who works both independently and for Qu Junktions. He is a member of filmmakers' co-op Bristol Experimental and Expanded Film, co-founder of underground art space The Brunswick Club, and a visiting lecturer at the RCA London. He has published various essays, as well as articles for *Kaleidoscope*, *The Wire* and Ibraaz.org.

Erik Davis is an author, scholar, and award-winning journalist based in San Francisco. He explores the 'cultures of consciousness' on his weekly podcast *Expanding Mind*, and recently earned his PhD in religious studies at Rice University. His next book, *High Weirdness: Drugs, Esoterica, and Visionary Experience in the Seventies* (Strange Attractor/MIT Press), will be published in 2019.

Kodwo Eshun is Lecturer in Contemporary Art Theory at Goldsmiths, University of London, Visiting Professor, Haut Ecole d'Art et Design, Genève, and co-founder of The Otolith Group.

Matthew Fuller is the author of books including *How to Sleep: The Art, Biology and Culture of Unconsciousness* (Bloomsbury, 2018), and is Professor of Cultural Studies at Goldsmiths, University of London.

Kristen Gallerneaux is the author of *High Static, Dead Lines: Sonic Spectres and the Object Hereafter* (Strange Attractor/MIT Press) and has written for the Barbican Center, *ARTnews*, and *The Quietus*. She is also Curator of Communications and Information Technology at the Henry Ford Museum in Detroit, Michigan, where she continues to build upon one of the largest historic technology collections in North America.

Lee Gamble is a UK-based sound designer, junglist, composer, and DJ. His flair for probing, warping, and dissecting stereotypical conceptions of electronic sound led to the release of the seminal *Diversions 1994–1996*, followed by *Dutch Tvashar Plumes*, *Koch*, and *Kuang* on PAN, and *Mnestic Pressure* and 2019's *In a Paraventral Scale* on Hyperdub, and the inception of UIQ (www.u-i-q.org), a platform dedicated to new voices in electronic music.

Agnès Gayraud is a French musician and philosopher born in 1979. She teaches theory at the Villa Arson (National Art School) in Nice. A former student of the Ecole Normale Supérieure (Ulm), she has made various contributions on current issues in Critical Theory, aesthetics, and modernity. Between musical practice (as La Féline) and critique (for the daily paper *Libération*), she recently published her first book about the aesthetics of recorded popular music: *Dialectique de la pop* (2018, La Philharmonie/La Découverte), considered as a major work on the subject.

Steve Goodman is a member of AUDINT. His book *Sonic Warfare: Sound, Affect and the Ecology of Fear* was published by MIT Press in 2009. He is founder of record label Hyperdub and sometimes responds to the name Kode9.

Olga Goriunova is co-author (with Matthew Fuller) of *Bleak Joys* (Minnesota University Press, forthcoming 2019), author of *Art Platforms* (Routledge, 2012), and editor of many collections. She is Reader at Royal Holloway, University of London.

Anna Greenspan is Assistant Professor of Contemporary Global Media at NYU Shanghai. Her research interests include urban Asia, emerging media, philosophy of technology, and Chinese modernity. Her latest book *Shanghai Future: Modernity Remade* was published by Oxford University Press in 2014. <http://www.annagreenspan.com>.

S. Ayesha Hameed's moving image, performance and written work explore contemporary borders and migration, and visual cultures of the Black Atlantic. Her projects *Black Atlantis* and *A Rough History (of the destruction of fingerprints)* have been performed and exhibited internationally. She is the co-editor of *Futures and Fictions* (Repeater, 2017), and is currently the Programme Leader for the MA in Contemporary Art Theory in the Department of Visual Cultures at Goldsmiths, University of London.

Tim Hecker is a Canadian composer and has produced a range of audio works for numerous labels including Kranky, Mille Plateaux and 4AD. His work has also included commissions for contemporary dance, film scores, sound installations, as well as various writings. He completed a PhD in Communication Studies and Art History at McGill University in 2013. He is currently based in Los Angeles.

Julian Henriques is Professor and Joint Head of the Department of Media and Communications at Goldsmiths, University of London. He previously ran the Film and Television Department at CARI-MAC, at the University of the West Indies in Kingston, Jamaica. He is the author of *Sonic Bodies. Reggae Sound Systems, Performance Techniques, and Ways of Knowing* (Continuum, 2011).

Toby Heys is a member of AUDINT, and is a reader in Digital Media and the head of research for the School of Digital Arts (SODA) at Manchester Metropolitan University. He has a cross-dis-ciplinary research and practice profile but his dominant focus revolves around the ways that frequencies are utilised by governments and industry to influence, manipulate and torture. His monograph on this topic, *Sound Pressure*, will be published by Rowman and Littlefield in 2019.

Eleni Ikoniadou is a member of AUDINT and Senior Tutor in Visual Communication at the Royal College of Art. Her research is situated at the intersection between computational culture, the-ory-fiction and audiovisual practice. Her latest monograph is *The Rhythmic Event: Art, Media, and the Sonic* (MIT Press, 2014). She is co-editor of the Media Philosophy series (Rowman & Littlefield International).

Amy Ireland is a theorist and experimental writer based in Melbourne, Australia. Her research focuses on questions of agency and technology in modernity, and she is a member of the tech-no-materialist trans-feminist collective, Laboria Cuboniks.

The Occulture (David Cecchetto, Marc Couroux, Ted Hiebert, Eldritch Priest, and Rebekah Sheldon) is an experimental theory collective investigating the esoteric imbrications of sound, affect, and hyperstition. Their collectively authored book—*Ludic Dreaming: How to Listen Away From Contemporary Technoculture* (Bloomsbury, 2017)—uses dreams as a method for examining the decussation of sound and contemporary technoculture.

Nicola Masciandaro is Professor of English at Brooklyn College (CUNY) and a specialist in medieval literature. He is the author of *On the Darkness of the Will* (Mimesis, 2018).

Ramona Naddaff teaches in the Rhetoric Department at the University of California, Berke-ley, where she is also director of the Doreen B. Townsend Humanities Center's Art of Writing program. Naddaff's publications include studies of ancient philosophy and literature, literary censorship theory, and postwar French theory. She is currently researching a new project of modern case studies in the censorship of music. Naddaff is co-founder and co-director of Zone Books.

Anthony Nine is a writer and artist from the UK, now based in Miami. His work explores intersecting themes of occultism, African Diaspora traditions, psychogeography, music and culture. He is the author of *Space Weather Report*, a colouring book account of the world of spirit, available from Revelore Press. His first novel *Dub Seance*, a story of lived magic set in London and New Orleans, is forthcoming.

Luciana Parisi is Reader in Cultural Theory, Chair of the PhD program in Cultural Studies, and co-director of the Digital Culture Unit at Goldsmiths, University of London. Her research focuses on cybernetics, information theory and computation, complexity and evolutionary theories, and the technocapitalist investment in artificial intelligence, biotechnology, and nanotechnology. She is the author of *Abstract Sex: Philosophy, Biotechnology and the Mutations of Desire* (Bloomsbury, 2004) and *Contagious Architecture: Computation, Aesthetics, and Space* (MIT Press, 2013). She is currently researching the history of automation and the philosophical consequences of logical thinking in machines.

Alina Popa cares for a place from which it is possible to have artistic consequences, without a total break-up of life and art, of the politics of production and the politics of the product, of oneself as subjectivity and oneself as performance, of the art piece and its conditions of possibility. She thus found herself at the border—between visual arts and contemporary dance, the white cube and the black box, writing and theory. She founded The Bureau of Melodramatic Research together with Irina Gheorghe, the Bezna series of publications, and, more recently, Unsorcery, Black Hyperbox, Artworlds with Florin Flueraș.

Paul Purgas is a London-based electronic musician, artist and curator. Originally trained as an architect he has presented performances and installations with various public institutions and festivals and is currently based at Somerset House. He is one half of the experimental music project Emptyset releasing material through Thrill Jockey in Chicago.

Georgina Rochefort is a US-based cryptanalyst who was mostly active from the late 1940s–mid 1980s, working for commercial and government organisations.

Steven Shaviro is DeRoy Professor of English at Wayne State University in Detroit. He works mostly these days on science fiction and on music videos. His recent books include *Discognition* (Repeater, 2016) and *Digital Music Videos* (Rutgers University Press, 2017).

Jenna Sutela works with words, sounds, and other living materials. Her installations and performances seek to identify and react to precarious social and material moments, often in relation to technology. Sutela's work has been presented at museums and art contexts internationally, including Guggenheim Bilbao, Museum of Contemporary Art Tokyo, and Serpentine Galleries.

Jonathan Sterne is James McGill Professor of Culture and Technology at McGill University. He is author of *MP3: The Meaning of a Format* (Duke, 2012), *The Audible Past: Cultural Origins of Sound Reproduction* (Duke, 2003); and numerous articles on media, technologies and the politics of culture. He is also editor of *The Sound Studies Reader* (Routledge, 2012) and co-editor of *The Participatory Condition in the Digital Age* (Minnesota, 2016). His current projects consider instruments and instrumentalities; mail by cruise missile; and the intersections of disability, technology and perception. His next book, tentatively titled *Tuning Time: Histories of Sound and Speed*, is co-authored with Mara Mills. <http://sterneworks.org>.

Eugene Thacker is the author of several books including *Infinite Resignation* (Repeater, 2018) and *In the Dust of This Planet* (Zero Books, 2011). He teaches at The New School in New York City.

Dave Tompkins has contributed to *The Wire*, *The New Yorker*, *New York Magazine*, and *Oxford American*. His first book, *How To Wreck A Nice Beach* (Melville House, 2010) is a history of the vocoder, from World War II to hip-hop. He is currently working on a natural history of Miami Bass.

Shelley Trower is Reader in the Department of English and Creative Writing at the University of Roehampton. Her publications include *Senses of Vibration* (Bloomsbury, 2012) and *Rocks of Nation* (Manchester University Press, 2015). Other projects include 'Memories of Fiction: An Oral History of Readers' Life Stories' (<http://www.memoriesoffiction.org>).

INDEX